Springer Tracts in Advanced Robotics

Volume 75

Javier Civera, Andrew J. Davison,
and José María Martínez Montiel

Structure from Motion Using the Extended Kalman Filter

Springer

Professor Bruno Siciliano, Dipartimento di Informatica e Sistemistica, Università di Napoli Federico II, Via Claudio 21, 80125 Napoli, Italy, E-mail: siciliano@unina.it

Professor Oussama Khatib, Artificial Intelligence Laboratory, Department of Computer Science, Stanford University, Stanford, CA 94305-9010, USA, E-mail: khatib@cs.stanford.edu

Professor Frans Groen, Department of Computer Science, Universiteit van Amsterdam, Kruislaan 403, 1098 SJ Amsterdam, The Netherlands, E-mail: groen@science.uva.nl

Authors

Dr. Javier Civera
Universidad de Zaragoza
Instituto Universitario de Investigación en
Ingeniería de Aragón (I3A)
Edificio Ada Byron
María de Luna, 1, 50018 Zaragoza, Spain
E-mail: jcivera@unizar.es

Dr. Andrew J. Davison
Imperial College London
Department of Computing
180 Queen's Gate
SW7 2AZ London, UK
E-mail: ajd@doc.ic.ac.uk

Dr. José María Martínez Montiel
Universidad de Zaragoza
Instituto Universitario de Investigación en
Ingeniería de Aragón (I3A)
Edificio Ada Byron
María de Luna, 1, 50018 Zaragoza, Spain
E-mail: josemari@unizar.es

ISBN 978-3-642-24833-7 e-ISBN 978-3-642-24834-4

DOI 10.1007/978-3-642-24834-4

Springer Tracts in Advanced Robotics ISSN 1610-7438

Library of Congress Control Number: 2011940008

Typeset & Cover Design: Scientific Publishing Services Pvt. Ltd., Chennai, India.

Printed on acid-free paper

5 4 3 2 1 0

springer.com

STAR (Springer Tracts in Advanced Robotics) has been promoted under the auspices of EURON (European Robotics Research Network)

To María, María Ángeles, Pedro José and Sergio.

To Lourdes, Rafael, Blanca and Adela.

To José María and Olga. To Virginia, Elena and Mikel

Foreword

Robotics is undergoing a major transformation in scope and dimension. From a largely dominant industrial focus, robotics is rapidly expanding into human environments and vigorously engaged in its new challenges. Interacting with, assisting, serving, and exploring with humans, the emerging robots will increasingly touch people and their lives.

Beyond its impact on physical robots, the body of knowledge robotics has produced is revealing a much wider range of applications reaching across diverse research areas and scientific disciplines, such as: biomechanics, haptics, neurosciences, virtual simulation, animation, surgery, and sensor networks among others. In return, the challenges of the new emerging areas are proving an abundant source of stimulation and insights for the field of robotics. It is indeed at the intersection of disciplines that the most striking advances happen.

The *Springer Tracts in Advanced Robotics (STAR)* is devoted to bringing to the research community the latest advances in the robotics field on the basis of their significance and quality. Through a wide and timely dissemination of critical research developments in robotics, our objective with this series is to promote more exchanges and collaborations among the researchers in the community and contribute to further advancements in this rapidly growing field.

The monograph written by Javier Civera, Andrew J. Davison and J.M.M. Montiel is devoted to the wide field of computer vision for robotics, which has been receiving a great deal of attention by the research community throughout the last three decades. The contents are focused on the development of models and methods for sequential Structure from Motion or monocular SLAM, where the projective nature of the camera is fitted under a Bayesian filtering framework. Point feature model, efficient and robust correspondences search, model selection for degenerate configurations and internal self-calibration are essential to the theoretical framework underlying the sequential estimation algorithms described in the book.

Sparse point-based 3D reconstruction and motion estimation from an image sequence are achieved assuming no knowledge over the scene nor the internal camera calibration parameters, making the resulting system to effectively work in real-time at video rate in typical room-sized scenarios.

STAR is proud to welcome yet another volume in the series dedicated to the popular area of SLAM!

Naples, Italy
August 2011

Bruno Siciliano
STAR Editor

Preface

The book you have in your hands is the result of the joint research of the authors from 2006 to 2009, that led to Javier Civera's PhD thesis. The contributions on it were first published in several international conferences and journals; specifically and in chronological order [Montiel *et al.* 2006, Civera *et al.* 2007a, Civera *et al.* 2007b, Civera *et al.* 2008a, Civera *et al.* 2008b, Civera *et al.* 2009b, Civera *et al.* 2009a, Civera *et al.* 2009c, Civera *et al.* 2010]. This book provides a unified view of the research spread on those papers, presenting a complete and robust system aimed to sequential 3D vision in real-time.

Along the years, many people have contributed in one or another manner to the results presented here. First of all, we are very grateful to our coauthors in the topics related with this book Óscar G. Grasa, Diana R. Bueno and Juan A. Magallón. Gabe Sibley and Joan Solà reviewed the first version of the PhD thesis manuscript. The jury of Javier Civera's PhD defence; Wolfram Burgard, J. D. Tardós, Patric Jensfelt, Miguel Ángel Salichs and Víctor Muñoz, contributed to this final version with very insightful comments and annotations. The expert eyes of Antonio Agudo, Óscar G. Grasa, Lina Paz y Pedro Piniés also reviewed the final version of this book and contibuted to polish some discussions and to find the latest and more difficult *bugs*. The authors are also extremely grateful to their respective research groups: the Robot Vision Group at Imperial College in London and the Robotics, Perception and Real-Time Group in the I3A (Aragón Institute on Engineering Research) in the University of Zaragoza in Spain. Hundreds of relevant conversations, the stimulating research atmosphere and the dedication at both groups are fertile breeding grounds that have played an essential role in our research. An exhaustive list of names would be an unfeasible task, but still we would like to express our gratitude to some of our colleagues for very interesting discussions: José Neira, Pedro Piniés, Lina Paz, Rubén Martínez-Cantín, Luis Montesano, Alejandro Mosteo, Ana Cristina Murillo, Iñaki Raño, Antonio Agudo, Margarita Chli, Hauke Strasdat, Ian Reid, Brian Williams and Cyrill Stachniss. Last but not least, we are in debt with our respective families, for their constant support.

The research results in this book were financially supported by the following sources: The Spanish Goverment research grants DPI2003-07986, DPI2006-13578,

DPI2009-07130 and PR2007-0427; regional grant DGA(CONAI+D)-CAI IT12-06; the European Union projects RAWSEEDS FP6-045144 and RoboEarth FP7-248942; and A. J. Davison's ERC Starting Grant 210346.

Zaragoza and London
July 2011

Javier Civera
Andrew J. Davison
J.M.M. Montiel

Contents

1 Introduction 1
1.1 Structure from Motion and Monocular SLAM 3
1.1.1 Structure from Motion 3
1.1.2 Monocular SLAM 3
1.1.3 Structure of the Problem 4
1.2 Background 8
1.3 Outline of the Book 9

2 Points at Infinity: Mosaics Using the Extended Kalman Filter 13
2.1 Introduction 13
2.1.1 Points at Infinity 13
2.1.2 Real-Time EKF-Based Mosaicing 14
2.2 Related Work 15
2.2.1 SLAM 15
2.2.2 Off-Line SFM vs. EKF SLAM 16
2.2.3 Image Mosaicing 17
2.3 Geometrical Modeling 19
2.3.1 Feature Model 19
2.3.2 Camera Motion Model 19
2.3.3 Measurement Model 20
2.4 Simultaneous Localization and Mapping 21
2.4.1 Matching 22
2.4.2 State Initialization 23
2.4.3 Feature Initialization and Deletion 23
2.5 Meshing and Mosaicing 24
2.5.1 Updating the Mesh 24
2.5.2 Tile Texture Mapping 25
2.6 Experimental Results 26
2.6.1 360° Pan and Cyclotorsion 26
2.6.2 Real-Time Mosaic Building 29
2.6.3 Processing Time 30
2.7 Discussion 31

3 Inverse Depth Parametrization 33
3.1 Introduction 33
3.1.1 Delayed and Undelayed Initialization 34
3.1.2 Points at Infinity 36
3.1.3 Inverse Depth Representation 36
3.1.4 Inverse Depth in Computer Vision and Tracking 37
3.2 State Vector Definition 39
3.2.1 Camera Motion 39
3.2.2 Euclidean XYZ Point Parametrization 39
3.2.3 Inverse Depth Point Parametrization 39
3.2.4 Full State Vector 40
3.3 Measurement Equation 40
3.4 Measurement Equation Linearity 41
3.4.1 Linearized Propagation of a Gaussian 42
3.4.2 Linearity of XYZ Parametrization 43
3.4.3 Linearity of Inverse Depth Parametrization 44
3.4.4 Depth vs. Inverse Depth Comparison 44
3.5 Feature Initialization 45
3.6 Switching from Inverse Depth to XYZ 47
3.6.1 Conversion from Inverse Depth to XYZ Coding 47
3.6.2 Linearity Index Threshold 47
3.7 Data Association 48
3.7.1 Patch Warping 49
3.7.2 Active Search 50
3.8 Experimental Results 51
3.8.1 Indoor Sequence 52
3.8.2 Real-Time Outdoor Sequence 53
3.8.3 Loop Closing Sequence 55
3.8.4 Simulation Analysis for Inverse Depth to XYZ Switching 58
3.8.5 Parametrization Switching with Real Images 60
3.8.6 Processing Time 62
3.9 Discussion 62

4 1-Point RANSAC 65
4.1 Introduction 65
4.2 Related Work 68
4.2.1 Random Sample Consensus (RANSAC) 68
4.2.2 Joint Compatibility Branch and Bound (JCBB) 69
4.2.3 Structure from Motion and Visual Odometry 71
4.2.4 Benchmarking 71
4.3 1-Point RANSAC Extended Kalman Filter Algorithm 72
4.3.1 EKF Prediction and Search for Individually Compatible Matches (lines 5–8) 72

4.3.2 1-Point Hypotheses Generation and Evaluation (lines 9–22) . . . 74
4.3.3 Partial Update with Low-Innovation Inliers (lines 23–24) . . . 75
4.3.4 Partial Update with High-Innovation Inliers (lines 25–35) . . 75
4.4 1-Point RANSAC Extended Kalman Filter from a Monocular Sequence Input . . . 76
4.4.1 State Vector Definition . . . 76
4.4.2 Dynamic Model . . . 77
4.4.3 Camera-Centered Estimation . . . 77
4.5 Experimental Results . . . 80
4.5.1 Benchmark Method for 6 DOF Camera Motion Estimation . . . 80
4.5.2 1-Point RANSAC . . . 81
4.5.3 Joint Compatibility Branch and Bound (JCBB) . . . 84
4.5.4 Trajectory Benchmarking against GPS . . . 85
4.5.5 Pure Monocular EKF-Based Estimation for Long Sequences . . . 89
4.5.6 Visual Odometry from a Monocular Sequence Plus Wheel Odometry . . . 90
4.6 Discussion . . . 93

5 Degenerate Camera Motions and Model Selection . . . 99
5.1 Introduction . . . 99
5.2 Bayesian Model Selection for Sequences . . . 101
5.3 Interacting Multiple Model . . . 103
5.4 Interacting Multiple Model Monocular SLAM . . . 104
5.5 Experimental Results . . . 105
5.5.1 Consistent Start Up Even with Rotation . . . 106
5.5.2 Low Risk of Spurious Matches due to Small Search Regions . . . 106
5.5.3 Camera Motion Model Identification . . . 109
5.5.4 Computational Cost Considerations . . . 110
5.6 Discussion . . . 110

6 Self-calibration . . . 111
6.1 Introduction . . . 111
6.2 Related Work . . . 112
6.3 Sum of Gaussians (SOG) Filter . . . 113
6.4 Self-calibration Using SOG Filtering . . . 115
6.4.1 State Vector Definition . . . 115
6.4.2 Pruning of Gaussians with Low Weight . . . 116
6.5 Experimental Results . . . 117
6.5.1 Indoor Sequence . . . 118
6.5.2 Loop-Closing Sequence . . . 119
6.6 Discussion . . . 122

7 Conclusions 123

Appendix

A Implementation Details 127
A.1 Extended Kalman Filter 127
A.2 Calibrated EKF-Based SfM 129
A.2.1 Dynamic Model and Derivatives 129
A.2.2 Measurement Model and Derivatives 131
A.2.3 Inverse Depth Point Feature Initialization and Derivatives 136
A.3 Uncalibrated EKF-Based SfM 139
A.3.1 Dynamic Model and Derivatives 139
A.3.2 Measurement Model and Derivatives 140
A.3.3 Inverse Depth Point Feature Initialization and Derivatives 143
A.4 Quaternion Normalization 146
A.5 Inverse Depth to Cartesian Parameterization Conversion 147

B Filter Tuning Understanding via Dimensional Analysis 149
B.1 Introduction 149
B.2 Monocular SLAM Estimation Process 150
B.3 Buckingham's Π Theorem Applied to Monocular SLAM 151
B.4 Dimensionless Monocular SLAM Model 151
B.5 Geometric Interpretation of the Dimensionless Parameters 152
B.6 Real Image Results 154
B.6.1 Dependence of Scene Scale on a Priori Parameters 154
B.6.2 Image Tuning in a Pure Rotation Sequence 155
B.6.3 The Same Image Tuning for Different Sequences 156
B.7 Discussion 157

References 159

Chapter 1
Introduction

Abstract. The fully automated 3D estimation of a scene and the 6 degrees of freedom camera motion using as the only input the images taken by the camera has been a long term aim in the computer vision community. The intense research in the latest decades has produced spectacular advances; the topic is already mature and most of its aspects are already well known. 3D vision has inmediate applications in many different fields, like robotics or augmented reality, and technological transfer is starting to be a reality. This book describes one of the first systems for sparse point-based 3D reconstruction and egomotion estimation from an image sequence; able to work in real-time at video rate in room-sized scenarios. This chapter introduces the main topics that will be detailed in the rest of the book, covering aspects like the point feature model, efficient and robust correspondences search, model selection for degenerate configurations and internal self-calibration.

One of the most brilliant quotes attributed to Albert Einstein says that you do not really understand something unless you can explain it to your grandmother. With that in mind, computer vision researchers should consider themselves rather lucky to be able to summarize the main aim of their discipline with a simple and understandable sentence like "making a computer see". On the other hand, the lexical simplicity of this objective hides a very complex reality which very often people are tricked into. Even relevant researchers of the field are said to have fallen into the trap: The anecdote that Marvin Minsky, Artificial Intelligence pioneer from MIT, assigned to solve the whole computer vision problem as a summer project to a degree student back in the sixties is an illustrative and well-known example [Hartley & Zisserman 2004].

The truth behind this apparent simplicity is that, although we all have a clear experience about what "to see" implies, the biological mechanisms of visual processing are yet not fully understood. And even if we knew it, we could also wonder if a machine needs –or will be able to run– a visual sensing similar to ours. This lack of a precise definition about what "to see" really means and needs have made of computer vision a diverse and fragmented discipline.

J. Civera et al.: Structure from Motion Using the Extended Kalman Filter, STAR 75, pp. 1–12.
springerlink.com

In spite of this, computer vision has experienced great advances since its appearance. Computers still cannot see, but nowadays most of them are able to use visual information in one or another sense. Also, cameras are becoming the dominant sensor modality in many fields, for example in the robotics research or the industry of videogames.

The general frame of the work is one of these uses of visual information, specifically how visual information can be processed to extract a sparse tridimensional estimation of the imaged scenario and the motion of the camera into it. Figure 1.1 shows a typical example of the estimation results in this book[1]. Figure 1.1(a) shows the image k in the video sequence, along with the tracked features. Figure 1.1(b) shows a 3D top view of the estimation results: the black triangle stands for the current pose of the camera, the black line is the trajectory up to the current frame, and the red ellipses represent the 3D 95% uncertainty region for the point features.

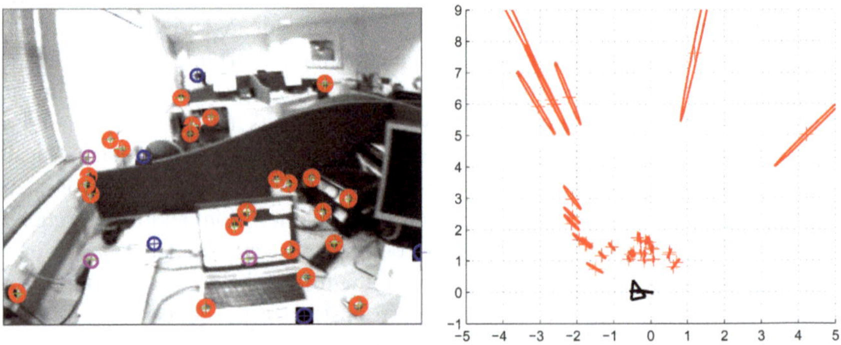

(a) Image from the input sequence and tracked features. Ellipses in the image are the regions where the correspondences are searched. Correspondences are plotted as green crosses.

(b) Top view of the 3D estimation results. The black triangle stands for the current camera pose, and the black line is the estimated trajectory up to the current frame. The estimated features are plotted as red crosses and their uncertainty as red ellipses.

Fig. 1.1. Example illustrating the usual EKF SfM or monocular SLAM processing.

The algorithms described in this book provide a theoretical framework to perform a sequential estimation of a camera motion and 3D scene from the only input of an image sequence and in real-time up to 30 frames per second. The contents of the book allow to perform such 3D estimation *out-of-the-box*; that is, for any sequence and any camera motion, assuming no knowledge over the scene nor the internal camera calibration parameters.

[1] Produced with the open source Matlab code in `http://webdiis.unizar.es/~jcivera/code/1p-ransac-ekf-monoslam.html`

This comes from the application of solid theoretical concepts, deeply rooted in multiple view geometry and probability theory. As another output of the application of a well-founded theory, also the length and the accuracy of the estimation are greatly improved; from waggling camera motion in indoors scenarios and sequences of around a minute to half-an-hour sequences of tens of thousands frames taken by a robot covering trajectories of hundreds of metres.

1.1 Structure from Motion and Monocular SLAM

1.1.1 Structure from Motion

Inside the computer vision field, Structure from Motion (SfM) is the line of research that, taking as the only input a set of image correspondences, seeks to infer in a totally automated manner the 3D structure of the scene viewed and the camera locations where the images where captured. SfM has been one of the most active areas of research for the latest three decades, reaching such a state of maturity that some of its algorithms have already climbed to the commercial application level [2d3 2011, Autosticht 2011, PhotoTourism 2011].

SfM origins can be traced back to the so-called photogrammetry, that since the second half of 19th century aimed to extract geometric information from images. Starting with a set of features manually identified by the user, photogrammetry makes use of non-linear optimization techniques known as Bundle Adjustment (BA) to minimize the reprojection error [Mikhail *et al.* 2001]. The research done by the computer vision community has been mostly oriented to achieve the complete automation of the problem and has produced remarkable progress in three aspects: first, the constraints imposed on the motion of the features in two –or three– images under the assumption of the rigidity of the scene have been formalized, even in the case of degenerate motion and uncalibrated camera [Hartley & Zisserman 2004]; second, intense research on salient feature detection and description with a high degree of invariance [Canny 1986, Harris & Stephens 1988, Lowe 2004]; and third, spurious rejection [Fischler & Bolles 1981, Rousseeuw & Leroy 1987]. The result is an automated way of robustly matching point and line features along images and estimate the geometric relations between pairs.

Based on these three achievements several methods have been proposed that, from a set of images of a scene, are able to estimate the three-dimensional structure and camera locations up to a projective transformation in the most general case of uncalibrated cameras. With some extra knowledge about the camera calibration, a metric solution up to scale can be obtained. This SfM estimation from constraints between pairs or triplets of images is usually followed by a Bundle Adjustment (BA) step [Triggs *et al.* 2000] that minimizes the reprojection error and refines this initial pairwise estimation into a globally consistent one.

1.1.2 Monocular SLAM

On the other hand, the estimation of the ego-motion of a mobile platform and its surroundings has also been tackled by the robotics community from a slightly different point of view. The so-called SLAM (standing for Simultaneous Localization and Mapping) [Durrant-Whyte & Bailey 2006, Bailey & Durrant-Whyte 2006] has been referred as one of the fundamental problems in mobile robotics research. The SLAM addresses the estimation of the motion of a mobile robot and a map of its surroundings from the data stream provided by one or several sensors. The first SLAM approaches made use of multiple sensors, e.g. laser [Castellanos *et al.* 1999, Newman *et al.* 2002], radar [Dissanayake *et al.* 2001], or sonar [Tardós *et al.* 2002]. Most of the times, wheel odometry measurements were also included in the estimation and sensorial fusion was also a recurrent topic [Castellanos *et al.* 2001]. As computer vision became mature enough, vision gained a predominant position in this sensor fusion schemes and even was used as the only sensorial input [Davison *et al.* 2007]. Monocular SLAM refers to the use of a monocular camera as the dominant sensor, and sometimes the only sensor, for performing SLAM.

Structure from Motion and monocular SLAM present another difference that is rather relevant for defining the purposes of this book. While SfM has dealt with the problem in its most general form –that is, for any kind of visual input–, monocular SLAM has focused in sequential approaches for the processing of video input. This comes as a consequence of the specificity of the robotic application; as the sensorial input in a robot comes naturally in the form of a stream. A mobile robot also needs sequential estimation capabilities: At every step the best possible estimation is required in order to insert it in the control loop, and hence batch processing does not make sense. This sequential constraint is not limited to the robotics applications: Augmented reality, for example, also needs a sequential estimation in order to coherently and smoothly insert a virtual character in every frame of a sequence and hence can make use of monocular SLAM algorithms [Klein & Murray 2009].

1.1.3 Structure of the Problem

Following the notation in [Strasdat *et al.* 2010a], both SLAM and SfM can be represented in terms of a Bayesian network. Figure 1.2 shows a simplified example that illustrates this. In the Bayesian network in this figure, the variables we want to estimate are: $\mathbf{x}_{C_k}$, containing a set of parameters related to the camera that captured image I_k –in the most general case, internal and external calibration parameters; and $\mathbf{y}_i$, containing a set of parameters modeling a 3D feature in the scene. The observed or measured variables are $\mathbf{z}_i^k$, that represent the projection of a feature i in an image I_k. The measurement model imposes constraints between the variables we want to estimate and the image measurements, that are the arrows in the net. As long as we have enough constraints, the variables $\mathbf{x}_{C_k}$ and $\mathbf{y}_i$ can be estimated up to a certain extent [Hartley & Zisserman 2004]. Notice that in the specific case of an image sequence input, extra constraints can be added from the variables of one camera $\mathbf{x}_{C_k}$ to the next one $\mathbf{x}_{C_{k+1}}$ by modeling the dynamics of the camera motion. In 1.2, this extra links that may appear in the formulation are plotted in dotted line.

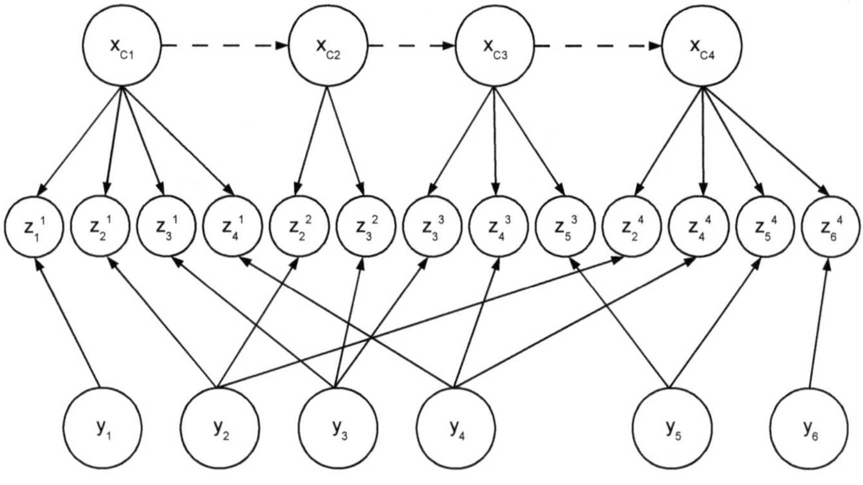

Fig. 1.2. Bayesian network for SfM.

For the sake of clarity, figure 1.3 models the previous example as a Markov random field. Here, the constraints imposed by the image measurements $\mathbf{z}_i^k$ and the motion model are represented by the links between the hidden variables. If one were to solve this Bayesian network offline, the standard approach would be to perform a global optimization over the whole network. Such global optimization in this specific 3D vision problem receives the name of Bundle Adjustment [Triggs *et al.* 2000]. In order to initialize the iterative Bundle Adjustment optimization, the correspondences between pairs or triplets of images provide the geometric constraints to calculate their relative motion [Hartley & Zisserman 2004, Nistér 2004].

The estimation of the 3D estructure and camera motion in a sequential manner can be modeled with a Markov random field very similar to the one in figure 1.2; but taking into account that the network grows at every step with the addition of new camera poses for each frame processed, new 3D features as the camera images new areas, and new tracked image features that link both. This new Markov random field is shown in figure 1.4. The aim in sequential SfM –or monocular SLAM– consists of propagating over all the hidden camera and feature variables every new image evidence. At the most basic level of local SLAM, two different approaches are mostly used nowadays:

- *Keyframe-based SLAM.* This first approach adapts the offline Bundle Adjustment to fit the sequential processing of a video sequence. The SLAM computation is separated into two parallel threads: first, one thread for high frame rate camera tracking, where the camera pose is estimated from the known features that are

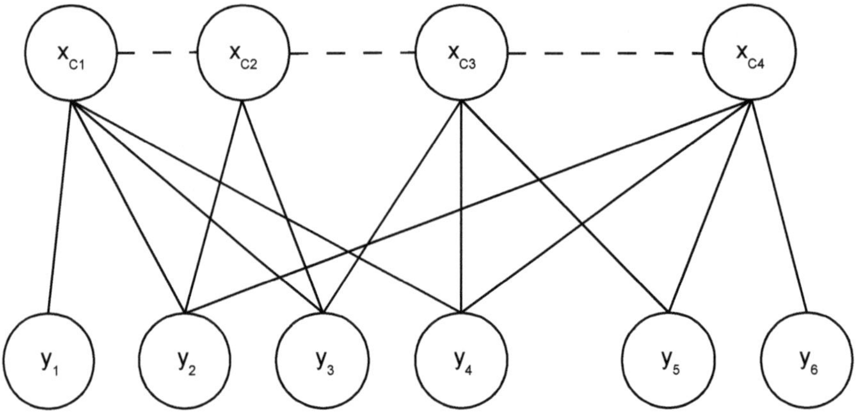

Fig. 1.3. Markov random field for SfM.

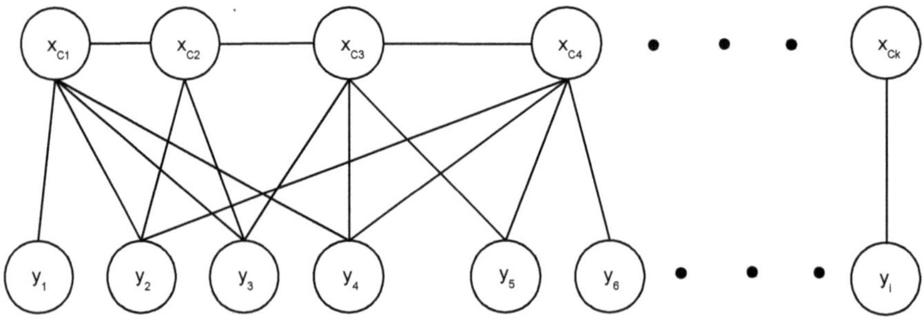

Fig. 1.4. Markov random field for sequential SfM or monocular SLAM.

already in the map. The second thread performs, at a much lower rate, a global optimization over the tracked features and a selected set of keyframes in the image sequence. Such keyframes can correspond to the whole sequence [Klein & Murray 2007, Klein & Murray 2008, Konolige & Agrawal 2008, Strasdat *et al.* 2010b] or a sliding window around the current frame [Nistér *et al.* 2006, Mouragnon *et al.* 2006, Konolige *et al.* 2007]. The tracked camera pose in the first thread serves as the initial seed for this iterative non-linear refinement in the second thread.

Figure 1.5 illustrates this first approach. Notice that in this simplified example the intermediate cameras $\mathbf{x}_{C_2}$ and $\mathbf{x}_{C_3}$ are not keyframes and hence are removed from the estimation. Notice also that the links between features $\mathbf{y}_2$, $\mathbf{y}_3$, $\mathbf{y}_4$, $\mathbf{y}_5$ and camera poses $\mathbf{x}_{C_2}$ and $\mathbf{x}_{C_3}$ have been also removed from the estimation and hence their information is lost.

- *Filtering-based SLAM.* Differently from keyframe optimization, filtering marginalize out the past poses and keep a joint probabilistic estimation over the current pose and the 3D map features. A filtering algorithm adds at every step

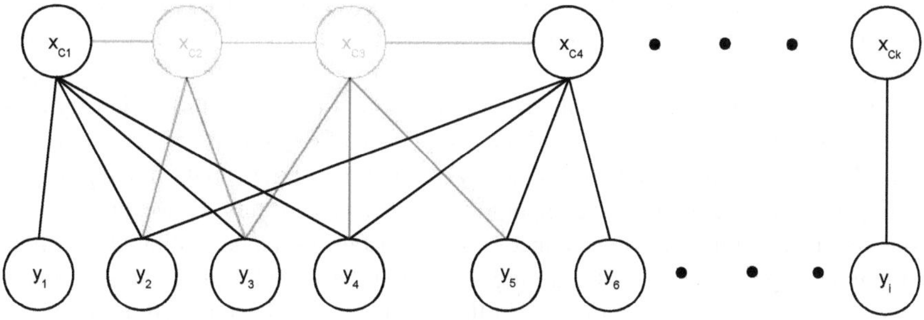

Fig. 1.5. Keyframe-based SLAM.

the current camera variables to the estimation, marginalizes out previous cameras and update the whole network based on the image measurements in the current frame.

Figure 1.6 illustrates this in the proposed example. Notice that the marginalization of a past pose introduces probabilistic links between every pair of features in the map that do not appear in the keyframe SLAM methods. The Markov random field in figure 1.6 presents less nodes than the keyframe-based one in figure 1.5, but on the other hand extra links have appeared between every pair of features. It is also worth remarking that filtering algorithms marginalize the links between past camera poses and features –that is, they integrate their information into the current state– instead of removing them like keyframe SLAM does.

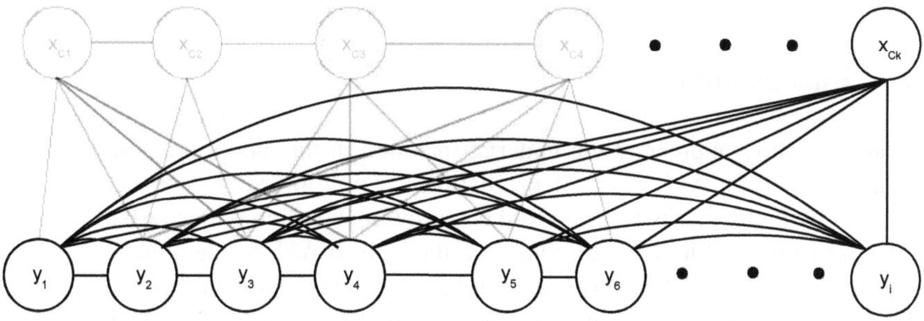

Fig. 1.6. Filtering-based SLAM.

The more sparse structure of the links in keyframe SLAM methods makes them less computationally demanding than filtering-based ones. It can be argued that they also process less information than filtering methods: in figure 1.5 only the correspondences in images $I_i; i = 1,4$, which are the links between camera parameters $\mathbf{x}_{C_1}$ and $\mathbf{x}_{C_4}$ and features $\mathbf{y}_i; i = 1,\ldots,6$ are considered in the estimation. A filtering algorithm is able to integrate the information in every image $I_i; i = 1,\ldots,4$ before marginalizing past camera parameters. Nevertheless, it has been demonstrated very recently [Strasdat *et al.* 2010a] that the ratio between the information gained by considering this extra cameras and the computational cost derived from a denser structure compared with a keyframe SLAM is small enough to be worth including them.

On the other hand, filtering algorithms maintain a joint probability density estimation over the current camera and map in a more efficient manner than keyframe ones. Most of the times, the high accuracy of the 3D visual estimation in a practical setting makes unnecesary the computation of such uncertainty. However, it is also true that the computation of such uncertainty distribution can be profitable in critical situations –e.g sequence bootstrapping, small number of tracked features or quasi-degenerate configurations [Frahm & Pollefeys 2006]—. Filtering methods could be advisable in those cases.

Structure from Motion under a Bayesian filtering framework, or visual SLAM, has been a relevant chapter on 3D estimation from images with examples as [Davison 2003] based on Extended Kalman Filter, [Eade & Drummond 2006] based on particle filters, [Holmes *et al.* 2008] based on Unscented Kalman Filter and [Eade & Drummond 2007] based on Information Filters. This book targets visual systems based on the Extended Kalman Filter (EKF); having the system in [Davison *et al.* 2007] as its baseline. Nevertheless, most of the concepts proposed can be extended from the EKF to the other filtering schemes. Although the Extended Kalman Filter is often accused of consistency problems due to its strong requirements for linearity and Gaussianity [Julier & Uhlmann 2001, Bailey *et al.* 2006, Castellanos *et al.* 2007]; it will be shown along the results of this book that those requirements hold for the Structure from Motion problem at a local scale and the EKF performance matches that of other methods. Further insight in this issue can also be read in the recent [Solà *et al.* 2011].

1.2 Background

In order to fully apprehend the material included in this book, the reader should be familiar with SLAM, Bayesian estimation and multiple view geometry. Regarding SLAM, the reader is referred to the survey in [Durrant-Whyte & Bailey 2006, Bailey & Durrant-Whyte 2006], that describes the SLAM problem and covers the essential algorithms. [Davison *et al.* 2007] describes the first monocular SLAM system demonstrating real-time performance and that we take as the starting point of this book. [Bar-Shalom *et al.* 2001, Gelb 1999] are excellent references for estimation and tracking from a general point of view. [Thrun *et al.* 2005] covers the main sequential estimation techniques applied in robotics including the Extended Kalman Filter; and hence is particularly advisable. The appendix A in this book contains a very detailed description of the Extended Kalman Filter formulation used along the book aimed to help the implementation of the presented results.

Regarding single and multiple view geometry, the first chapters of [Hartley & Zisserman 2004] introduce projective parameterizations, that will be discussed in the first chapters of this document, and the camera model that we use along the book. Regarding salient point detection and description, although a wide array of detectors and descriptors with a high degree of invariance are available for feature extraction and matching [Mikolajczyk *et al.* 2005, Mikolajczyk & Schmid 2005], the results presented in this book make use of the Harris corner detector [Harris & Stephens 1988] and plain normalized cross-correlation [Brown *et al.* 2003] between warped image patches [Hartley & Zisserman 2004] for matching. The maintenance of the probabilistic estimation over camera and features makes possible to warp planar patches according to the relative motion between frames, making unnecesary a high degree of invariance. For some of the results in the book, the FAST detector [Rosten & Drummond 2005] has also been used, reducing the feature extraction cost without noticeable performance improvement nor degradation.

1.3 Outline of the Book

The main aim of this book is then to develop models and methods for sequential SfM or monocular SLAM; fitting the projective nature of the camera under a Bayesian filtering framework. In this section, the specific topics of the book are introduced in more detail.

- **Chaper 2: Points at Infinity. Mosaics using the Extended Kalman Filter**
  In Structure from Motion it is well known that cameras, as projective sensors, are able to image very distant points, theoretically even points at infinity. Several illustrative examples can be taken from our daily lives: For example, we are able to see the stars even when they are several million light years away. Homogeneous coordinates have provided an algebraic manner to manipulate very distant points in SfM without numerical issues. Although distant features do not provide information for estimating the camera translation, they have proved to be very useful to estimate its orientation. Following the previous example, we cannot estimate our position in the earth looking at the stars, but we can extract very precisely directions –where the North is– by looking at them.
  Although the notation *close* and *distant* points may be the most intuitive one, the accuracy of the estimation in multiview geometry is actually governed by the *parallax angle*. The parallax angle is the angle formed by the two projection rays that goes from a point feature to the optical centers of two cameras. The bigger this angle, the more accurately the depth of the point can be estimated. Low values for this angle may be caused by distant features, but also by small translation between two images.
  Early visual filtering estimation techniques [Ayache & Faugeras 1989, Matthies *et al.* 1989, Broida *et al.* 1990, Azarbayejani & Pentland 1995, Chiuso *et al.* 2002, Davison 2003] had limitations with these low-parallax points. Hence, they were losing an important source for orientation information and limiting the use of cameras in outdoors scenarios where these type of features are rather common. This chapter discusses the value of such low-parallax points for camera orientation estimation. The extreme case of a camera

undergoing pure rotational motion –and hence zero-parallax points– is presented and processed using a filtering algorithm able to estimate camera rotation and point directions in real-time at 30 Hz. Loop closures of 360° are achieved as a proof of consistency.
As an application of this, a mosaicing application is built on top of the estimated backbone map. This mosaicing algorithm directly inherits the advantages of the EKF processing, being the first mosaicing technique that presents real-time drift-free spherical mosaicing results for camera rotations of 360°.

- **Chaper 3: Inverse Depth Parameterization**
The step that naturally follows camera rotation and zero-parallax point estimation is a unified parametrization for low and high parallax features, allowing to estimate camera rotation and translation. The key concept of the parametrization proposed in this chapter is the direct coding of the inverse depth of the features relative to the camera locations from which they were first viewed. This inverse depth parametrization offers two key improvements with respect to the standard Euclidean one. First, as homogeneous coordinates, it is able to model zero-parallax points. And second, the addition of the initial camera position improves the degree of linearity of the projection equation, which is a crucial requirement for the Extended Kalman Filter.
The inverse depth point feature model solves in an elegant manner what was called in monocular SLAM *the initialization problem*: Previous point feature models were unable to be inserted from the frame they were first seen in the general estimation framework due to the unobservability of the depth along the projection ray. The usual approach was to delay this initialization until the feature depth converged to a value. Features initialized in such delayed manner would not contribute to the estimation until they show parallax enough. Notice that, in the common case of zero-parallax features, they would never be added to the estimation and hence their information would be lost. Inverse depth feature initialization is undelayed in the sense that even distant features are immediately used to improve camera motion estimates, acting initially as bearing references but not permanently labeled as such.
The inverse depth parametrization remains well behaved for features at all stages of SLAM processing, but has the computational drawback that each point is represented by a six dimensional state vector as opposed to the standard three of a Euclidean XYZ representation. We also show in this chapter that once the depth estimate of a feature is sufficiently accurate, its representation can be safely converted to the Euclidean XYZ form, and propose a linearity index which allows automatic detection and conversion to maintain maximum efficiency — only low parallax features need be maintained in inverse depth form for long periods.
- **Chaper 4: 1-point RANSAC**
This chapter has a double objective: first, it is aimed to illustrate for the first time how filtering-based visual SLAM methods, without neither submapping nor loop closure capabilities, can reach an accuracy and trajectory length comparable to keyframe methods. Specifically, a camera-centered Extended Kalman Filter is used here to process a monocular sequence as the only input (and also combined

with wheel odometry), with 6DOF motion estimated. Also in this chapter features are kept "alive" in the filter while visible as the camera explores forward and are deleted from the state once they go out of the field of view. In a few words, the EKF operates in a "visual odometry" mode [Nistér *et al.* 2006].
"Forgetting" the map permits an increase in the number of tracked features per frame from tens to around 1–2 hundred. While improving the accuracy of the estimation, it makes computationally infeasible the exhaustive Branch and Bound search performed by standard JCBB for match outlier rejection. As the second contribution that overcomes this problem, we present here a RANSAC-like algorithm that exploits the probabilistic prediction of the filter. This use of prior information makes it possible to reduce the size of the minimal data subset to instantiate a hypothesis to the minimum possible of 1 point, increasing the efficiency of the outlier rejection stage by several orders of magnitude.
Experimental results from real image sequences covering trajectories of hundreds of meters are presented and compared against RTK GPS ground truth. Estimation errors are about 1% of the trajectory for trajectories up to 650 metres.

- **Chaper 5: Degenerate Camera Motions and Model Selection**
Degenerate camera motions have a capital importance from a practical point of view of implementing real systems. For example, if a camera is attached to a mobile robot, there will be large periods when the robot is stopped. Pure rotation motion is also very frequent in industrial robotic arms. Even in hand-held camera motion estimation, there will be periods where the camera is almost still, and rotations are more easily performed than large translations –particularly in outdoor environments.
Any estimation algorithm modeling a general camera motion in any of the above situations will fail. What happens here is that the image noise incorrectly fits the extra parameters of the model. This problem is referenced in the SfM literature, where model selection schemes are used when degenerate motion may be encountered. This schemes discriminate models based on two terms: a term based on the reprojection error, which basically discards simplistic models; and a penalty term based on the complexity of the model that avoids selecting overparameterized models.
What we propose here is a model selection scheme that computes probabilities over models, based on research on model selection carried out by the tracking community. In a probability-based scheme, simplistic methods will receive low probabilities, but also overparameterized ones as their probability distribution function expands over unnecessary dimensions. Ad-hoc penalty terms for complex models are not needed in the proposed probability-driven model selection.
- **Chaper 6: Self-calibration**
The Structure from Motion methods were developed assuming a minimum or even null knowledge neither about the camera nor the scene. We have already commented in previous paragraph that model selection algorithms were developed to cover *every* possible type of scene and camera geometric configuration. The geometric constraints that correspondences should hold have been formalized even in the case of any additional information than the images; that is, for

cameras with unknown and possibly varying calibration parameters. Using some reasonable additional information a metric reconstruction, egomotion estimation and internal self-calibration can be estimated up to a scale factor.
While this is a well-known result, monocular SLAM and sequential SfM have mostly used precalibrated cameras [Davison *et al.* 2007, Klein & Murray 2008, Konolige & Agrawal 2008, Mouragnon *et al.* 2009]. The authors believe that self-calibration would improve the usability of the SLAM algorithms. For certain applications, like augmented reality for endoscopic surgery [Grasa *et al.* 2011], self-calibration is a must: you cannot expect wasting the precious time of a medical team in a tedious calibration of an endoscope.
As a step towards this *out-of-the-box* SLAM, this chapter presents an algorithm for monocular SLAM or sequential SfM from an uncalibrated image sequence. Camera internal calibration, egomotion and 3D scene structure are estimated from an image sequence without any prior information.

- **Appendix A: Implementation Details**
This first appendix aims at providing every implementation detail that can be helpful for the reproduction of the algorithms in this book. Along the first seven chapters the emphasis has been in the concepts and experimental proofs; and some of the details have been hidden for the sake of a more clear presentation. Particularly, this chapter details the computation of the Jacobians for the dynamic and measurement models used along the book; which are needed by the EKF processing.
- **Appendix B: Filter Tuning Understanding via Dimensional Analysis**
As already said, it is a well known fact that in the best of the cases the Structure from Motion or pure monocular SLAM estimation can only provide a geometric estimation up to a scale factor. Nevertheless, in the formulation adopted in this book, the parameters in the state vector are modeled using metric units. More specifically, the filter tuning parameters (e.g., the acceleration noise) are the ones set in metric units and act as the metric priors for the 3D estimation. The scale of the estimation in every monocular experiment of the book is then meaningless, as it is unobservable, and comes as a result of the metric priors introduced in the filter.
The geometric priors that induce the scale of our experiments are detailed in this appendix. More importantly, this appendix shows a dimensional analysis of the SfM problem that allows: 1) the identification of the relevant magnitudes of the problem; and 2) the proposal of a new dimensionless formulation that separates the real scale of the estimation from the monocular SLAM and allows to represent the estimated parameters in terms of dimensionless length ratios and angles.

Chapter 2
Points at Infinity: Mosaics Using the Extended Kalman Filter

Abstract. This chapter introduces the use of zero-parallax points on filtering-based Structure from Motion (SfM) or monocular SLAM. A geometric model is proposed for the estimation of an accurate camera rotation and the directions of a set of tracked features using an Extended Kalman Filter. On top of that, it is proposed a sequential mosaicing algorithm able to build drift-free, consistent spherical mosaics in real-time, automatically and seamlessly even when previously viewed parts of the scene are re-visited. This method represents a significant advance on previous mosaicing techniques which either require an expensive global optimization or which run sequentially in real-time but use local alignment of nearby images and ultimately drift.

2.1 Introduction

2.1.1 Points at Infinity

In SfM, the well-known concept of a point at infinity is a feature which exhibits no parallax during camera motion due to its extreme depth. A star for instance would be observed at the same image location by a camera which translated through many kilometers pointed up at the sky without rotating. Such a feature cannot be used for estimating the camera translation but is a perfect bearing reference for estimating its rotation. Homogeneous coordinates, used in SfM, allow an explicit representation of points at infinity, and they have proven to play an important role during off-line structure and motion estimation.

This chapter presents a real-time EKF filtering algorithm for consistently estimating the motion of a rotating camera observing such points at infinity. This algorithm introduces the use of infinity points in filtering SfM approaches, stepping

J. Civera et al.: Structure from Motion Using the Extended Kalman Filter, STAR 75, pp. 13–32.
springerlink.com

forward to the use of those algorithms in outdoors environments where low-parallax features are very common. This step is culminated in next chapter with a unified parameterization for close and distant –even infinity– points. Here, the accuracy and consistency of the estimated camera rotation and feature map is shown in 360° loop closures experiments using real-image sequences.

2.1.2 Real-Time EKF-Based Mosaicing

The second contribution of this chapter is the use of a consistent map of point features as the basis for a real-time, drift-free mosaicing system. This is the first algorithm which can build drift-free mosaics over the whole viewsphere in seamless real-time: no backtracking, batch optimization or learning phase is required; and arbitrarily long image sequences can be handled without slow-down.

Mosaicing involves accurately aligning a set of overlapping images based on correspondences, normally between automatically-detected salient features. While mosaicing does not require full 3D camera motion and scene structure to be estimated, it can be considered as a reduced SfM or visual SLAM problem since it has the key requirements of simultaneously solving for camera pose (in this case rotation) and feature locations (in this case directions without a depth coordinate). Many different approaches to mosaicing have been published, being the most relevant for this chapter summarized in section 2.2.3. All are based on pair-wise image matching to estimate local camera motion. Those which aim to operate sequentially and in real-time simply concatenate multiple motion estimates and are prone to drift as errors accumulate. Those whose goal is globally consistent mosaics have a final off-line optimization step which re-adjusts the motion estimates to bring them into consensus.

As a key difference from previous works, image alignment is based here on a filtering scheme. Under this approach, it is built a persistent, *probabilistic* representation of the state of the sensor and scene map which evolves in response to motion and new sensor measurements. Specifically, it is used an Extended Kalman Filter (EKF) with a state vector consisting of stacked parameters representing the 3D orientation and angular velocity of the camera and the directions (i.e. viewsphere coordinates, since no depth can be estimated for the scene points) of a set of automatically acquired features, none of which need to be known in advance.

As a brief summary of the complete mosaicing algorithm, we take first the image stream from a rotating camera, build an efficient, persistent SLAM map of infinite points — *directions* mapped onto the unit sphere — and use these as the anchor points of a triangular mesh, built sequentially as the map eventually covers the whole viewsphere. Every triangle in the mesh is an elastic tile where scene texture is accumulated to form a mosaic. As each new image arrives, the probabilistic map of infinite points is updated in response to new measurements and all the texture tiles are re-warped accordingly. So, every measurement of a point potentially improves *the whole mosaic*, even parts not currently observed by the camera thanks to probabilistic knowledge of the correlations between estimates of different features. This attribute is especially valuable when a loop is closed because the whole map benefits from a large correction, removing drift.

The following chapter is organised as follows: After a literature review and comparison between off-line and sequential approaches in section 2.2; section 2.3 covers the essentials of the EKF estimation of infinite points and camera rotation. Section 2.4 describes how features are added to or removed from the map. Section 2.5 is devoted to the mesh for the mosaic given a map of feature directions. Finally experimental results and a brief discussion are presented in sections 2.6 and 2.7.

2.2 Related Work

2.2.1 SLAM

The standard approach to SLAM is to use a single Gaussian state vector to represent the sensor and feature estimates; and to update this with the Extended Kalman Filter (EKF). This approach was called the 'stochastic map' when proposed initially by Smith and Cheeseman in [Smith & Cheeseman 1986a] and has been widely used in mobile robotics with a range of different sensors; odometry, laser range finders, sonar, and vision among others (e.g. [Castellanos & Tardós 1999, Feder *et al.* 1999, Ortín *et al.* 2003]). This amounts to a rigorous Bayesian solution in the case that the sensor and motion characteristics of the sensor platform in question are governed by linear processes with Gaussian uncertainty profiles — conditions which are closely enough approximated in real-world systems for this approach to be practical in most cases, and in particular for small-scale mapping.

Visual sensing has been relatively slow to come to the forefront of robotic SLAM research. Davison and Murray [Davison & Murray 1998] implemented a real-time system where a 3D map of visual template landmarks was build and observed using fixating stereo vision. Castellanos et al. [Castellanos *et al.* 1994] built a 2D SLAM system combining straight segments from monocular vision and odometry, and trinocular straight segments and odometry. In [Davison 2003] Davison demonstrated 3D SLAM using monocular vision as the only sensor, also using a smooth motion model for the camera to take the place of odometry this system exhibited unprecedented demonstrable real time performance for general indoors scenes observed with a low cost hand-held camera. After this seminal work, many other interesting 3D SLAM systems which rely only on visual sensing started to emerge (e.g. [Kim & Sukkarieh 2003, Jung & Lacroix 2003, Eustice *et al.* 2005]).

2.2.2 Off-Line SFM vs. EKF SLAM

In order to directly compare the off-line approaches to mosaicing with our sequential one, we will focus on methods which rely on discrete feature matching. The off-line approaches generally match features between images in a pairwise fashion (usually between temporal neighbours) and then perform a final global bundle adjustment optimization to achieve good drift-free solutions.

In our sequential approach, scene structure — a set of selected points we call a map — and the camera motion are estimated by iterating the following steps:

1. Predict the locations of *all* the map features in the next image, along with a gated search region for each. The information from all previous images is implicitly accumulated in this state-based prediction and this leads to tight search regions.
2. Use template matching to exhaustively search for the feature match only within its search region.
3. Update estimates of the locations of *all* the mapped features and the current camera location.
4. Map maintenance, adding new features when new scene areas are explored and marginalizing out features predicted to be visible but persistently not matched.

The resulting sparse set of map features, autonomously selected and matched has — when compared with 'first match and then optimize' approaches — the following desirable qualities for producing consistent mosaics:

Long tracks — Only features persistently observed when predicted to be visible are kept in the map, and are allowed to build up immunity to future deletion. Highly visible and identifiable features tend to live on in this process of 'survival of the fittest'.

Loop closing tracks — When the camera revisits an area of the scene previously observed, it has the natural ability to identify and re-observe 'old', loop closing, features seamlessly.

To achieve the highest accuracy in every update of a scene map after matching image features, the update step should ideally be done using an iterative non-linear bundle adjustment optimization, but this would be prohibitively expensive to be run at every step. Instead of that, the EKF can serve as a sequential approximation to bundle adjustment. Update by bundle adjustment after processing every single image means a non linear optimization for *all camera locations* and all scene features, so processing long image sequences results in an increasing number of camera locations and hence an increasing dimension for the bundle adjustment of $(3m_k + 2n_k)$, where m_k is the number of images, and n_k is the number of scene points. Also, calculating search regions requires the inversion of a matrix of dimension $3m_k + 2n_k$ to compute the estimation covariance.

To compare the EKF with bundle adjustment, it should be considered that, as stated in [Triggs *et al.* 2000], the EKF is a sequential approximation to bundle adjustment where:

1. A motion model is included to relate one camera location with the next.
2. The EKF is just bundle adjustment's first half-iteration because only the most recent camera location is computed. The estimated state is reduced, subsuming all historic camera location estimates in the feature covariance matrix. The estimated state, dimension $(7+2n_k)$, is composed of the last camera pose and all map features. In our model the camera state vector has dimension 7: an orientation quaternion and 3D angular velocity.
3. At each step, information about the previous camera pose is marginalized out. In doing so, linearization for previous camera poses is not computed as in the optimal solution, and hence linearization errors will remain in the covariance matrix. However, mosaicing with a rotating camera is a very constrained, highly linear problem, so the results that can be obtained in real-time are highly accurate, only falling a little short of the result a full optimization would produce.

2.2.3 Image Mosaicing

Mosaicing is the process of stitching together data from a number of images, usually taken from a rotating camera or from a translating camera observing a plane, in order to create a composite image which covers a larger field of view than the individual views. While a variety of approaches have been published for matching up sets of overlapping images with a high degree of accuracy, all have relied on off-line optimization to achieve global consistency. Other previous methods which operate sequentially in a pair-wise manner and in real-time suffer from the accumulation of drift.

Mosaic building requires estimates of the relative rotations of the camera when each image was captured. Classically, the computation of such estimates has been addressed as an off-line computation, using pair-wise image matching to estimate local alignment and then global optimization to ensure consistency. The goal has normally been to produce visually pleasing panoramic images and therefore after alignment blending algorithms are applied to achieve homogeneous intensity distributions across the mosaics, smoothing over image joins. We focus here only on the alignment part of the process, achieving registration results of quality comparable to off-line approaches but with the advantage of sequential, real-time performance. Blending or other algorithms to improve the aesthetic appearance of mosaics could be added to our approach straightforwardly, potentially also running in real-time.

[Szeliski & Shum 1997] presented impressive spherical mosaics built from video sequences, explicitly recognizing the problem of closing a loop when building full panoramas (from sequences consisting of a single looped pan movement). However, their method needed manual detection of loop closing frames. The non-probabilistic approach meant that the misalignment detected during loop closing was simply evenly distributed around the orientation estimates along the sequence.

[Sawhney *et al.* 1998] tackled sequences with more complicated zig-zag pan motions, requiring a more general consideration of matching frames which were temporally distant as loops of various sizes are encountered. Having first performed pairwise matching of consecutive images they could estimate an approximate location for each image in the sequence. This was followed by an iterative hypotheses verification heuristic, applied to detect matching among geometrically close but temporally distant images to find the topology of the camera motion, and then finally global optimization. Both [Szeliski & Shum 1997] and [Sawhney *et al.* 1998] use whole image intensities without feature detection for the optimization.

[Capel & Zisserman 1998] proposed methods based on matching discrete features; RANSAC is applied to detect outlier-free sets of pairwise matches among the images. A global bundle adjustment optimization produces the final mosaic, achieving super-resolution. [de Agapito *et al.* 2001] also used robust feature matching to perform self-calibration and produced mosaics from sequences from a rotating and zooming camera.

[Brown 2003] considered the problem of mosaicing sets of widely separated, uncalibrated still images. Their method used SIFT features to perform wide-baseline matching among the images, and automatically align them into a panorama, optimizing over the whole set for global consistency using bundle adjustment.

The recent work of Steedly et al. [Steedly *et al.* 2005] is perhaps closest to the presented approach because it explicitly considers questions of computational cost in efficiently building mosaics from long video sequences (on the order of a thousand frames), though not arriving at real-time performance. The key to efficient processing in their system is the use of automatically assigned key-frames throughout the sequence — a set of images which roughly span the whole mosaic. Each frame in the sequence is matched against the nearest keyframes as well as against its temporal neighbors. The idea of building a persistent map of keyframes as the backbone of the mosaic can be thought of as very similar to the keyframe methods described in section 1.1 [Klein & Murray 2007, Klein & Murray 2008]. In fact, very recently, keyframe methods have been explicitly applied to mosaicing showing impressive real-time performance [Lovegrove & Davison 2010].

To our knowledge, up to [Civera *et al.* 2009b, Lovegrove & Davison 2010], mosaicing algorithms which truly operate in real-time have been much more limited in scope. Several authors have shown that the straightforward approach of real-time frame to frame image matching can produce mosaics formed by simply concatenating local alignment estimates. Marks et al. [Marks *et al.* 1995] presented a real-time system for mosaicing underwater images using correlation-based image alignment, and Morimoto and Chellappa a system based on point feature matching which estimated frame-to-frame rotation at 10Hz [Morimoto & Chellappa 1997]. It should be noted that while Morimoto and Chellappa used the EKF in their approach, the state vector contained only camera orientation parameters and not the locations of feature points as in our SLAM method.

The clear limitation of such approaches to real-time mosaicing is that inevitable small errors in frame to frame alignment estimates accumulate to lead to misalignments which become especially clear when the camera trajectory loops back on itself — a situation our SLAM approach can cope with seamlessly.

Some recent approaches have attempted to achieve global consistency in real-time by other means — Kim and Hong [Kim & Hong 2006] demonstrated sequential real time mosaic building by performing 'frame to mosaic' matching and global optimization at each step. However, this approach is limited to small-scale mosaics because the lack of a probabilistic treatment means that the cost of optimization will rise over time as the camera continues to explore. Zhu et al. [Zhu *et al.* 2006] on the other hand combine a frame to frame alignment technique with an explicit check on whether the current image matches to the first image of the sequence, detection of which leads to the correction of accumulated drift. Again, the method is not probabilistic and works only in the special case of simple panning motions.

2.3 Geometrical Modeling

We start the exposition of our method by explaining the mathematical models used for features, camera motion and the measurement process, before proceeding in section 2.4 to the EKF formulation.

2.3.1 Feature Model

Feature points are modeled by storing both geometric and photometric information (see Figure 2.1). Geometrically, the direction for feature i relative to a world frame W is parameterized as an angular azimuth/elevation pair:

$$\mathbf{y}_i^W = \begin{pmatrix} \theta_i \ \phi_i \end{pmatrix}^\top . \tag{2.1}$$

This corresponds to the following cartesian position $\mathbf{m}_i^W$ of the point on the unit sphere:

$$\mathbf{m}_i^W = \begin{pmatrix} \cos\phi_i \sin\theta_i & -\sin\phi_i & \cos\phi_i \cos\theta_i \end{pmatrix}^\top . \tag{2.2}$$

To represent each infinite point photometrically, a texture patch of fixed size is extracted and stored when the point is imaged for the first time. This patch is used for correlation-based matching.

2.3.2 Camera Motion Model

It is assumed that the camera translation is small compared with actual feature depths — true for any camera on a tripod, or well approximated by a camera rotated in the hand outdoors. The real-world camera dynamics are modeled as a smooth angular motion: specifically a 'constant angular velocity model', which states that at each time-step an unknown angular acceleration impulse drawn from a zero-mean Gaussian distribution is received by the camera. The camera state vector is:

$$\mathbf{x}_C = \begin{pmatrix} \mathbf{q}^{WC} \\ \omega^C \end{pmatrix}, \tag{2.3}$$

where ω^C is the angular velocity in the camera reference frame and $\mathbf{q}^{WC}$ is a quaternion defining the orientation with respect to a world reference frame W. See Figure 2.1 for further illustration on this model.

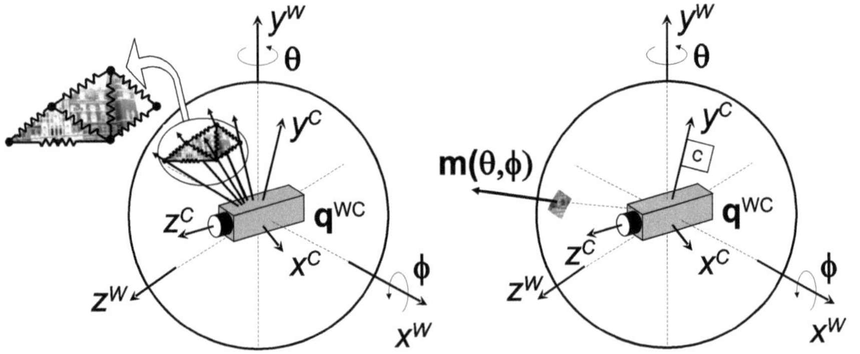

Fig. 2.1. Left, An elastic textured triangular mesh is built over a map of scene directions over the unit sphere. Right a scene feature, $\mathbf{y}_i^W$, stamped with its texture patch.

At every time-step k, an angular acceleration α^C having zero mean and fixed diagonal 'process noise' covariance matrix $\mathbf{P}_\alpha$ is assumed to affect the camera's motion. Therefore at each processing step of duration Δt the camera receives an impulse of angular velocity: $\Omega^C = \alpha^C \Delta t$. The state update equation is then:

$$\mathbf{x}_{C_{k+1}} = \begin{pmatrix} \mathbf{q}_{k+1}^{WC} \\ \omega_{k+1}^C \end{pmatrix} = \mathbf{f}\left(\mathbf{x}_{C_k}, \mathbf{n}\right) = \begin{pmatrix} \mathbf{q}_k^{WC} \times \mathbf{q}\left(\left(\omega_k^C + \Omega^C\right)\Delta t\right) \\ \omega_k^C + \Omega^C \end{pmatrix} \tag{2.4}$$

2.3.3 *Measurement Model*

We consider first the projection of infinite points in a standard perspective camera. The camera orientation $\mathbf{q}^{WC}$ in the state vector defines the rotation matrix $\mathbf{R}^{CW}$. So the coordinates of a point $\mathbf{y}^W = \begin{pmatrix} \theta & \phi \end{pmatrix}^\top$ on the unit sphere $\mathbf{m}$ expressed in frame C are:

$$\mathbf{m}^C = \mathbf{R}^{CW}\mathbf{m}^W = \mathbf{R}^{CW}\begin{pmatrix} \cos\phi\sin\theta & -\sin\phi & \cos\phi\cos\theta \end{pmatrix}^\top \tag{2.5}$$

The image coordinates where the point is imaged are obtained applying the pinhole camera model to $\mathbf{m}^C$:

$$\begin{pmatrix} u_u \\ v_u \end{pmatrix} = \begin{pmatrix} C_x - \alpha_x \frac{m_x^C}{m_z^C} \\ C_y - \alpha_y \frac{m_y^C}{m_z^C} \end{pmatrix} = \begin{pmatrix} C_x - \frac{f}{d_x} \frac{m_x^C}{m_z^C} \\ C_y - \frac{f}{d_y} \frac{m_y^C}{m_z^C} \end{pmatrix}, \tag{2.6}$$

where $(C_x \ C_y)^\top$ define the camera's principal point and α_x and α_y are the focal length in the horizontal and vertical image directions in pixel units. f is the focal length in metric units and d_x and d_y define the size of a pixel in metric units.

Finally, a distortion model has to be applied to deal with real camera lenses. In this work we have used the standard two parameter distortion model from photogrammetry [Mikhail *et al.* 2001], which is described next.

To recover the ideal projective undistorted coordinates $\mathbf{h}_u = (u_u, v_u)^\top$, from the actually distorted ones gathered by the camera, $\mathbf{h}_d = (u_d, v_d)^\top$, the following is applied:

$$\mathbf{h}_u = \begin{pmatrix} C_x + (u_d - C_x)\left(1 + \kappa_1 r_d^2 + \kappa_2 r_d^4\right) \\ C_y + (v_d - C_y)\left(1 + \kappa_1 r_d^2 + \kappa_2 r_d^4\right) \end{pmatrix}$$
$$r_d = \sqrt{(d_x (u_d - C_x))^2 + (d_y (v_d - C_y))^2} \tag{2.7}$$

Where κ_1 and κ_2 are the radial distortion coefficients.

To compute the distorted coordinates from the undistorted:

$$\mathbf{h}_d = \begin{pmatrix} C_x + \frac{(u_u - C_x)}{\left(1 + \kappa_1 r_d^2 + \kappa_2 r_d^4\right)} \\ C_y + \frac{(v_u - C_y)}{\left(1 + \kappa_1 r_d^2 + \kappa_2 r_d^4\right)} \end{pmatrix} \tag{2.8}$$

$$r_u = r_d \left(1 + \kappa_1 r_d^2 + \kappa_2 r_d^4\right) \tag{2.9}$$

$$r_u = \sqrt{(d_x (u_u - C_x))^2 + (d_y (v_u - C_y))^2} \tag{2.10}$$

r_u is readily computed computed from 2.10, but r_d has to be numerically solved from 2.9, e.g using Newton-Raphson, hence 2.8 can be used to compute the distorted point.

2.4 Simultaneous Localization and Mapping

The algorithm to estimate camera rotation and scene point directions follows the standard EKF loop [Bar-Shalom & Fortmann 1988] of prediction based on the motion model (equation 2.4), and measurement (equations 2.5 to 2.10). All the estimated variables (the camera state $\mathbf{x}_C$ and all the estimated feature directions $\mathbf{y}_i$) are stacked in a single state vector $\mathbf{x} = \left(\mathbf{x}_C^\top \ \mathbf{y}_1^\top \ \ldots \ \mathbf{y}_i^\top \ \ldots \ \mathbf{y}_n^\top\right)^\top$ with corresponding covariance $\mathbf{P}$ representing Gaussian-distributed uncertainty. Crucial to the method is the usage of the measurement prediction to actively guide matching. Next we explain in detail the matching process, the initialization of system state and feature initialization and deletion.

2.4.1 *Matching*

Every predicted measurement of a feature in the map, $\hat{\mathbf{h}}_i$, and its corresponding innovation covariance, $\mathbf{S}_i$, define an gated elliptical acceptance image region where the feature should lie with high probability. In our experiments, defining acceptance regions at 95% probability typically produce ellipses of diameter 10–20 pixels.

The first time a feature is observed, we store both a texture patch and the current camera orientation. When that feature is later selected for measurement after camera movement, its predicted texture patch from the current camera orientation is synthesized via a warping of the original patch. This permits an efficient matching of features from any camera orientation without the need for invariant feature descriptors. Figure 2.2 shows an example of the stored and predicted patches. Further details of the warping algorithm under general camera motion can be found in section 3.7.1.

Search for a feature during measurement is carried out by calculating a normalized correlation score for every possible patch position lying within the search region. The position with the highest score $\mathbf{z}_i$, provided that the match score is above a threshold, is considered to be feature i correspondence at this step.

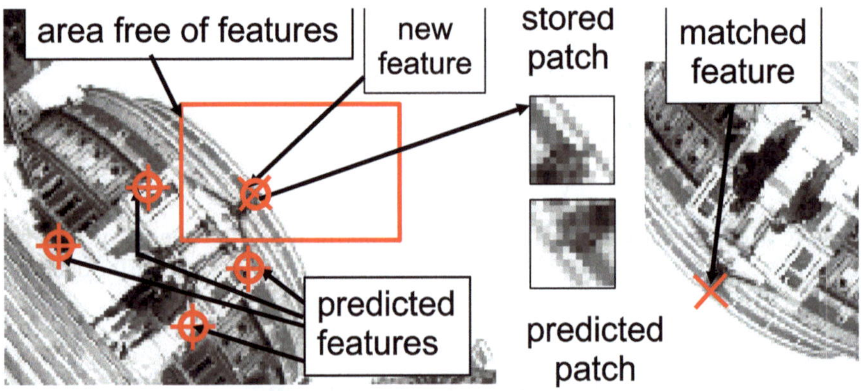

Fig. 2.2. New features initialization and patch prediction.

2.4.2 State Initialization

We initialize the state of the camera with zero rotation — its initial pose defines the world coordinate frame, and therefore we also assign zero uncertainty to initial orientation in the covariance matrix. The angular velocity estimate is also initialized at zero, but a high value is assigned to σ_ω, in our case $\sqrt{2}\,\frac{rad}{sec}$, in order to deal with an initial unknown velocity. This is a remarkable system characteristic: the map can be initialized from a camera which is already rotating. In fact in the experiments, the camera was already rotating when tracking commenced.

2.4.3 Feature Initialization and Deletion

When a new image is obtained, if the number of features predicted to be visible inside the image goes below a threshold, in our case around 15, a new feature is initialized. A rectangular area without features is selected randomly in the image, and searched for a single salient point by applying the Harris detector [Harris & Stephens 1988]. Figure 2.2 shows an initialization example.

This simple rule means that at the start of tracking the field of view is quickly populated with features which tend to be well-spaced and covering the image. As the camera rotates and some features go out of view, it will then demand that new features are initialized — but if regions of the viewsphere are revisited, new features will not be added to the map since old features (still held in the state vector) are simply re-observed.

When an initial measurement of a new feature $\mathbf{z}_j$ is obtained, the state vector is expanded with the new feature estimate $\hat{\mathbf{y}}_j$. Defining $\mathbf{R}_j$ the image measurement noise covariance, the covariance matrix is expanded as follows:

$$\mathbf{P}_{new} = \mathbf{J}\begin{pmatrix} \mathbf{P} & 0 \\ 0 & \mathbf{R}_j \end{pmatrix}\mathbf{J}^\top$$

$$\mathbf{J} = \begin{pmatrix} \mathbf{I}\,\mathbf{0} \\ \hline \mathbf{J}_1 \end{pmatrix}, \quad \mathbf{J}_1 = \left(\frac{\partial \mathbf{y}_j}{\partial \mathbf{x}_C}, \mathbf{0}, \ldots, \mathbf{0}, \frac{\partial \mathbf{y}_j}{\partial \mathbf{z}_j}\right).$$

The backprojection function from an initial image measurement $\mathbf{z}_j$ to a 3D ray $\mathbf{y}_j$ is further elaborated in the appendix A, along with its derivatives.

Features with a low successes/attempts ratio in matching — in practice 0.5 — are deleted from the map if at least 10 matches have been attempted. This simple map maintenance mechanism allows deletion of non-trackable features — for example those detected on moving objects or people. Non persistent static scene features (for instance caused by reflections) are also removed.

2.5 Meshing and Mosaicing

At any step k we have available a map of infinite points:

$$\mathscr{Y}^k = \left\{\mathbf{y}_i^W\right\}, \quad i = 1 \ldots n_k \,. \tag{2.11}$$

After processing every image, the location estimate of *every* point in the map is updated and hence changed. Additionally, on each step some map points might be deleted or added as new.

The mosaics we build are made up of a set of textured elastic triangular tiles attached to the map of infinite points. A mesh triangle T_j is defined as:

$$T_j = \left\{j_1, j_2, j_3, \mathrm{TX}_j\right\} \,, \tag{2.12}$$

where $\{j_1, j_2, j_3,\}$ identify the map points to which the triangle is attached and TX_j defines the triangle texture. The triangle is termed as *elastic* because the location estimates of the points to which it is attached are updated at every step, and hence the triangle and texture are deformed accordingly.

The triangular mesh mosaic at step k is defined as:

$$\mathscr{M}^k = \left\{T_j^k\right\}, \quad j = 1 \ldots m_k \,. \tag{2.13}$$

2.5.1 Updating the Mesh

As the map is updated at every step, the mosaic has to be updated sequentially as well. The mosaic update consists of updating the elastic triangles, and potential deletion and creation of triangles.

Figure 2.3 summarizes the algorithm to sequentially update the mosaic $\mathscr{M}^{k-1}$ into $\mathscr{M}^k$. After processing image k, the map $\mathscr{Y}^k$ is available. A spherical Delaunay triangulation $\mathscr{D}^k$ for the map $\mathscr{Y}^k$ points is computed using Renka's [Renka 1997] algorithm:

Input data: $\mathscr{M}^{k-1}$, $\mathscr{Y}^k$, image at step k.
Output data: $\mathscr{M}^k$.
Algorithm:
1.- $\mathscr{D}^k = \left\{D_l^k\right\}$, Spherical Delaunay triangulation.
2.- $\mathscr{M}^k$ is a subset of $\mathscr{D}^k$. Every D_l^k is classified as:

Elastically updated triangle: D_l^k already in $\mathscr{M}^{k-1}$, but not in $\mathscr{M}^k$. If visible in current image, the tile texture can be updated.
New and visible: D_l^k not in $\mathscr{M}^{k-1}$ and the visible in image k. D_l^k is included in $\mathscr{M}^k$. Texture is gathered from image k.
New but not visible: D_l^k not in $\mathscr{M}^{k-1}$ but not fully visible in image k. It is not added to $\mathscr{M}^k$.

Fig. 2.3. Triangular mosaic update algorithm.

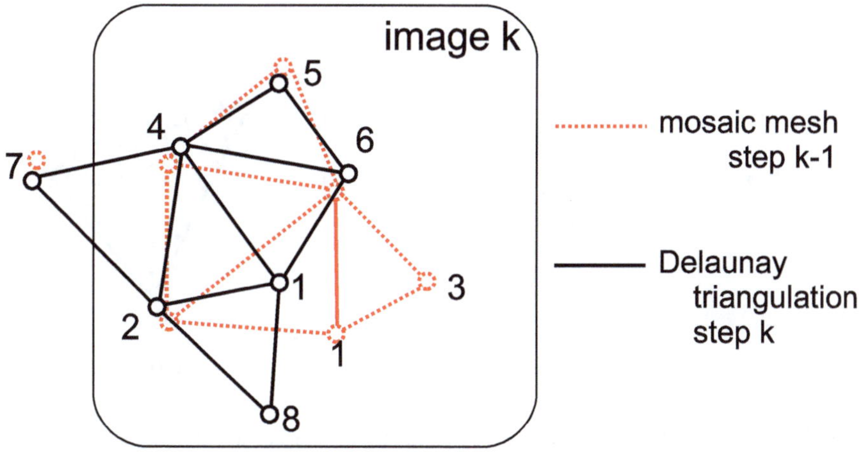

Fig. 2.4. Mesh update example. Triangle $\{456\}$ is elastically updated. $\{128\}$ new and visible: created because of new map point 8. $\{247\}$ new not visible; the triangle was not in $\mathscr{M}^{k-1}$, but it is not totally visible in image k. $\{136\}$ is deleted as map point 3 is removed. Points 1, 2, 4 and 6 are kept in the map but a significant change in the estimated position of point 1 has caused the triangulation to 'flip', so, $\{126\},\{246\}$ are deleted, while $\{124\},\{146\}$ are new and visible.

$$\mathscr{D}^k = \left\{ D_l^k \right\}, \tag{2.14}$$

where every triangle $D_l^k = \{l_1, l_2, l_3\}$ is defined by the corresponding 3 map features. The complexity of the triangulation is $O(n_k \log n_k)$ where n_k is the number of map features. The triangles which will be included in $\mathscr{M}^k$ are a subset of the triangles in the full triangulation $\mathscr{D}^k$. Every triangle D_l^k in $\mathscr{D}^k$ is classified to determine its inclusion in $\mathscr{M}^k$ mosaic according to the algorithm described in Figure 2.3: triangles are either carried over from the previous mosaic or newly created if texture can be immediately captured from the current image. Notice that triangles in $\mathscr{M}^{k-1}$ but not in D_l^k (due to a change in mesh topology) are deleted. Figure 2.4 illustrates with an example the different cases in the triangular mesh mosaic update.

2.5.2 Tile Texture Mapping

The three points defining a triangular tile are positions on a unit sphere. The texture to attach to each tile is taken from the region in the image between projections of these three vertices. In a simple approach to mosaic building, the triangles could be approximated as planar, and the mosaic surface a rough polyhedral. However better results can be achieved if the triangular tile is subdivided into smaller triangles that are backprojected over the spherical mosaic surface. Additionally, the camera

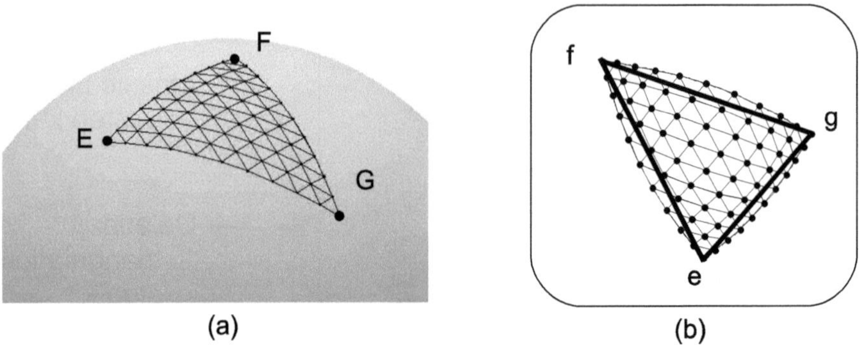

Fig. 2.5. (a) Triangular tile defined by map points E,F,G, meshed as subtriangles represented over the unit sphere. (b) Shows the backprojection in the image; notice how the subtriangles also compensate the radial distortion.

radial distortion is better compensated by the subtriangles. Figure 2.5 illustrates the improvement due to the division of a triangular tile into subtriangles.

2.6 Experimental Results

We present experiments demonstrating sequential mosaic building using images acquired with a low cost Unibrain IEEE1394 camera with a 90° field of view and 320 × 240 monochrome resolution at 30 fps. The first experiment shows the result from a sequence taken from a hand-held camera performing a 360° pan and cyclotorsion rotation — this early experiment performed offline in a Matlab implementation. The second experiment shows results obtained in real-time in a full C++ implementation with the more sophisticated sub-triangle mosaicing method. Both sequences were challenging because there were pedestrians walking around, and the camera's automatic exposure control introduced a great deal of change in the image contrast and brightness in response to the natural illumination conditions.

2.6.1 360° Pan and Cyclotorsion

The hand-held camera was turned right around about 1.5 times about a vertical axis, so that the originally-viewed part of the scene came back into view in 'loop closure'. Care was taken to ensure small translation, but relatively large angular accelerations were permitted and the camera rotated significantly about both pan and cyclotorsion axes.

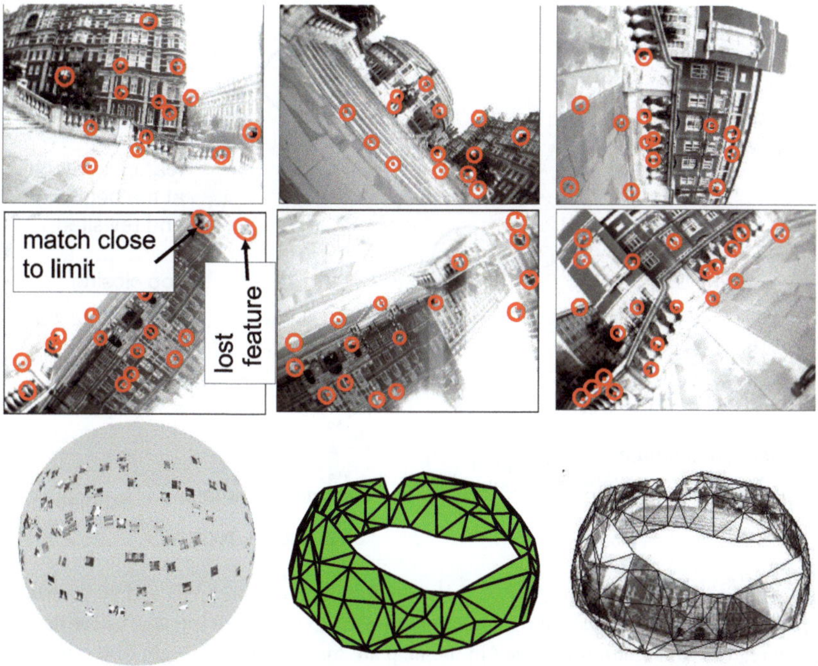

Fig. 2.6. 360° pan and cyclotorsion. Top left: first image in sequence. The first image on the second row is at the loop closure point; notice the challenging illumination conditions. Third row: unit sphere with feature patches, triangular mesh and simple texture.

Figure 2.6 shows selected frames showing the search region for every predicted feature and the matched observations. The frames we show are at the beginning of the sequence, at loop closure and during the camera's second lap. At the loop closure, we can see that of the first two re-visited features, one was not matched immediately and the other was detected very close to the limit of its search region. However, in the following frames most of the re-visited features were correctly matched despite the challenging illumination changes. It should be noticed that the loop is closed seamlessly by the normal sequential prediction-match-update process, without need any additional steps — no back-tracking or optimization. The mosaic after processing all the images is also shown.

We believe that the seamless way that loop closing is achieved is in itself a very strong indication of the achieved angular estimation accuracy with this algorithm. The predictions of feature positions just before redetection at loop closure only differ from their true values by around 6 pixels (corresponding to less than 2 degrees) when the camera has moved through a full 360°. After loop closing and the correction this forces on the map, this error is significantly improved. On the second lap, where the rotation of the camera takes it past parts of the scene already mapped, previously observed features are effortlessly matched in their predicted positions, the map having settled into a satisfyingly consistent state.

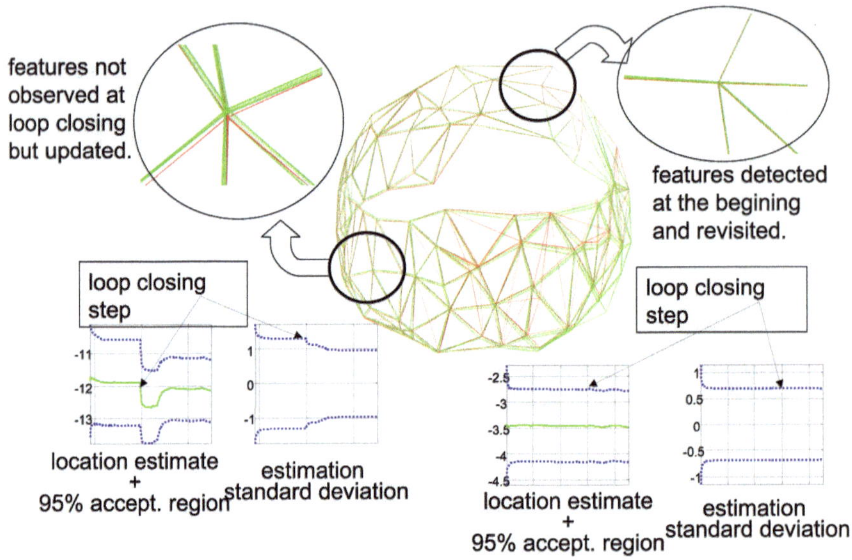

Fig. 2.7. Superimposed meshes for 90 frames before loop closure (red) and 90 frames after loop closure (green). The part of the mesh opposite the loop closure is magnified at the left of the figure. The first features detected in the map are magnified at the right of the figure. We also show the camera elevation estimation history along with its standard deviation; notice the uncertainty reduction at loop closure.

It is worth noting the effect that loop closing has on the mesh, and hence on the mosaic. Figure 2.7 displays superimposed all of the meshes for the 90 steps before the loop closure (we plot one in every five steps in red), along with meshes for the 90 steps after the loop closure (plotted in green). Every sequential step improves our estimates of the locations of all the map features, but in a loop closure step the improvement is much greater. As the initial camera orientation is chosen to be the base reference, the first features in the map are very accurately located (with respect to the first camera frame). These features' locations are hardly modified by the loop closure, but all other parts of the mesh are updated significantly. We particularly draw attention to an area of the mesh 180° opposite the loop closure point, where the feature estimates and therefore the mesh are noticeably modified in the update corresponding to loop closure thanks to the correlation information held in covariance matrix, despite the fact that these features are not observed in any nearby time-step.

Figure 2.7 also shows graphs of the estimate history for the two magnified features, along with the standard deviation in their elevation angles. The feature far from the loop close area clearly shows a loop closing correction and covariance reduction. The feature observed at the start of the sequence shows almost no loop closing effect because it was observed when the camera location had low uncertainty.

2.6.2 *Real-Time Mosaic Building*

A version of the system has been implemented in C++ achieving real time performance at 320×240 pixels resolution, 30 frames/second. We show results from a 360° pan rotation and map size of around a hundred features.

Figure 2.9 shows the evolution of the mosaic. We focus on the texture alignment at loop closure. Figures 2.9(g) and 2.9(i) display two magnified mosaic views close to the loop closure. In each magnified view, the left-hand part of the mosaic seen got its texture from a frame at the beginning of the sequence while the right area got texture from a frame after the loop closure (frames nearly a thousand images apart). The excellent texture alignment achieved is an indicator of the advantages of our sequential SLAM approach to mosaic building. No blending technique has been applied to reduce the seam effects.

Figure 2.8 shows an example of robustness with respect to moving objects. A feature was intialized on the skirt of a walking pedestrian. As the feature corresponds to a moving object, after some time it is no longer matched *inside* the acceptance region, and finally it is deleted from the map. It is also important to remark that no outlier rejection technique –like RANSAC or JCBB– was used here to detect the false match, being its rejection only due to the restricted search inside the gated 95% probability region.

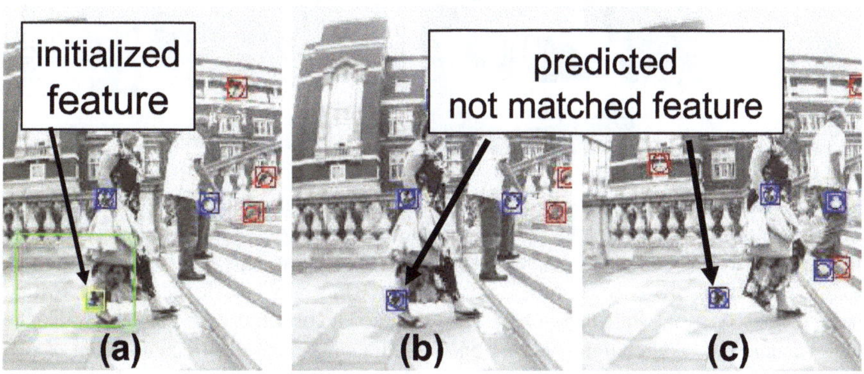

Fig. 2.8. (a) Feature initalized on a moving object. (b) after 1 and (c) 10 frames, the feature is no longer matched because it is outside its acceptance region. Eventually this non matching feature is deleted from the map.

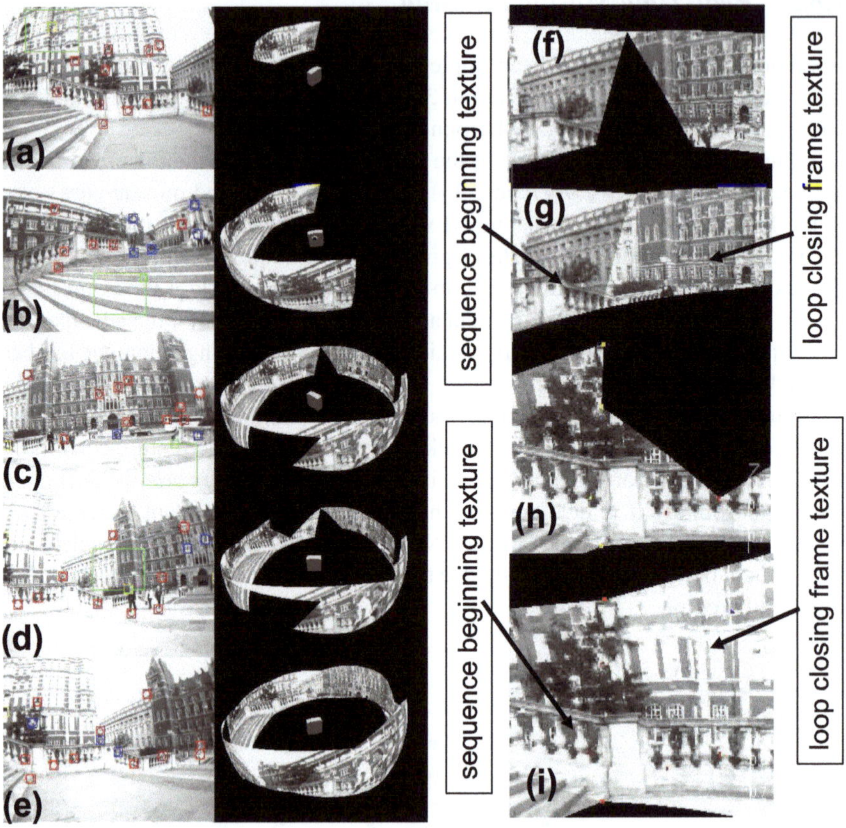

Fig. 2.9. Real-time mosaicing with loop closure. (a), (b) sequence start; (c) the frame after the first loop closing match; (d) a few frames after loop closing (e) the mosaic after almost two full laps. (f) (magnification of (c)) and (g) show two consecutive steps after loop closing; (g) is a close-up view of two adjacent mosaic areas. In (g) the left-most mosaic tiles got their texture early in the sequence, while those on the right obtained texture after loop closing. Notice the high accuracy in texture alignment; the line highlights the seam. (h) (magnification of (d)) and (i) show two consecutive steps after several loop closing matches. (i) is a close-up view of two adjacent mosaic areas with textures taken from early and loop closing frames; again notice the alignment quality. A line highlights the seam —otherwise difficult to observe.

2.6.3 Processing Time

Real-time experiments were run on a 1.8 GHz Pentium laptop with OpenGL accelerated graphics card. In a typical run, we might have: a) 70 map features, implying a state vector dimension of $7 + 70 \times 2 = 147$. b) 15 features measured per frame, implying a measurement vector dimension of $15 \times 2 = 30$. c) 30 fps, so 33.3 ms available for processing. d) Process noise with standard deviation $4 rad\ s^{-2}$ modeling expected angular accelerations.

Under these conditions, an approximate breakdown of typical processing time 21.5ms per frame is as follows: a)Image acquisition 1 ms. b) EKF prediction 3 ms. c) Image matching 2 ms. d) EKF update 10 ms. e), f) Delaunay triangulation 0.5 ms. g) Mosaic update 5ms.

The remaining time processing is used for the graphics functions, scheduled at low priority, so a graphics refresh might take place every two or three processed images.

It should be noticed that the computational complexity of the EKF updates is of order $O(N^2)$, where N is the number of map features. The Delaunay triangulation step has complexity of order $O(N \log N)$, while updating the triangles of the mosaic currently has order $O(N^2)$ (since each triangle is compared with every other one, though it should be possible to improve this). In our algorithm the EKF update dominates the computational cost. It should be noticed that within the bounds of a spherical mosaicing problem (where the whole viewsphere can be mapped with on the order of 100 features) the complexity is well within practical limits for a standard laptop.

2.7 Discussion

A mosaic built from an elastic triangular mesh sequentially built over a EKF SLAM map of points at infinity inherits the advantages of the sequential SLAM approach: probabilistic prior knowledge management through the sequence, sequential updating, real-time performance and consistent loop closing.

The experimental results presented using real images from a low cost camera display the validity of the approach in a challenging real scene with jittery hand-held camera movement, moving people and changing illumination conditions. Real-time seamless loop closing is demonstrated, removing all the drift from rotation estimation and allowing arbitrarily long sequences of rotations to be stitched into mosaics: the camera could rotate all day and estimates would not drift from the original coordinate frame as long as the high-quality features of the persistent map could still be observed. We think that as well as its direct application to mosaicing, this work shows in general the power of the SLAM framework for processing image sequences and its ability, when compared with off-line methods, to efficiently extract important information: pure sequential processing, long track matches, and loop closing matches.

EKF-based mosaicing is intended to offer a contribution when compared with other mosaicing approaches. Full 3D mosaicing in real-time is still some way off, so we have focused on building 2D mosaics from sequences with homography geometry — in this piece of work, specifically from a purely rotating camera, though we believe that our method could be straightforwardly modified to also cope with mosaicing a plane observed by a rotating and translating camera.

We believe that there are many applications which open up with real-time mosaicing — in any situation where the goal is to build a living mosaic which is built up instantly in reaction to camera motion our approach will be useful. This mosaic can find application especially in augmented reality because it provides a real-time link between the camera images and real scene points. An interesting application that is partly based in the algorithms described in this and next chapter is the one described at [Grasa *et al.* 2009b, Grasa *et al.* 2009a]. There, texture patches are attached to a backbone of 3D points in order to improve visualization of cavities inside human body, helping the medical team in laparoscopic surgery.

Another interesting possibility real-time mosaicing gives is that a user could control the rotation of the camera while looking at the growing mosaic in order to extend and improve it actively. In future work, we intend to look at such issues as real-time super-resolution, where we envisage a mosaic sharpening before a user's eyes thanks to the quality of repeatable rotation registration.

Chapter 3
Inverse Depth Parametrization

Abstract. This chapter presents a parametrization for point features within monocular SLAM which permits efficient and accurate representation of uncertainty during undelayed initialisation and beyond, all within the standard EKF (Extended Kalman Filter). The key concepts are direct parametrization of the inverse depth of features relative to the camera locations from which they were first viewed and the addition of this camera position at initialization to the state vector. The parametrization fits the projective nature of the camera in the sense that is able to code the infinite depth case. Also, the projection equation holds the high degree of linearity required by the Extended Kalman Filter. The chapter also shows that once the depth estimate of a feature is sufficiently accurate, its representation can safely be converted to the Euclidean XYZ form, and proposes a linearity index which allows automatic detection and conversion to maintain maximum efficiency –only low parallax features need be maintained in inverse depth form for long periods.

3.1 Introduction

A monocular camera is a projective sensor which measures the bearing of image features. Given an image sequence of a rigid 3D scene taken from a moving camera, it is now well known that it is possible to compute both the scene structure and the camera motion up to a scale factor. To infer the 3D position of each feature, the moving camera must observe it repeatedly, each time capturing a ray of light from the feature to its optic center. The measured angle between the captured rays from different viewpoints is the feature's parallax –this is what allows its depth to be estimated.

In off-line Structure from Motion (SfM) solutions from the computer vision literature (e.g. [Fitzgibbon & Zisserman 1998, Pollefeys *et al.* 1999]), motion and

J. Civera et al.: Structure from Motion Using the Extended Kalman Filter, STAR 75, pp. 33–63.
springerlink.com

structure are estimated from an image sequence by first applying robust feature matching between pairs or other short overlapping sets of images to estimate relative motion. An optimization procedure then iteratively refines global camera location and scene feature position estimates such that features project as closely as possible to their measured image positions (bundle adjustment). Recently, work in the spirit of these methods but with "sliding window" processing and refinement rather than global optimization has produced impressive real-time Visual Odometry results when applied to stereo [Nistér *et al.* 2006] and monocular sequences [Mouragnon *et al.* 2009].

An alternative approach to achieving real-time motion and structure estimation are on-line visual SLAM (Simultaneous Localization And Mapping) approaches which use a probabilistic filtering approach to sequentially update estimates of the positions of features (the map) and the current location of the camera. These SLAM methods have different strengths and weaknesses than visual odometry, being able to build consistent and drift-free global maps but with a bounded number of mapped features. The core single Extended Kalman Filter (EKF) SLAM technique, previously proven in multi-sensor robotic applications, was first applied successfully to real-time monocular camera tracking [Davison *et al.* 2007] in a system which built sparse room-sized maps at 30Hz. The sequential approach to the Structure from Motion problem has been also the subject of intense research in the computer vision community, being [Ayache & Faugeras 1989, Matthies *et al.* 1989, Broida *et al.* 1990, Azarbayejani & Pentland 1995, Chiuso *et al.* 2002] some of the most representative works.

A significant limitation of these early approaches, however, was that they could only make use of features which were close to the camera relative to its distance of translation, and therefore exhibited significant parallax during motion. The problem was in initialising uncertain depth estimates for distant features: in the straightforward Euclidean XYZ feature parametrization adopted, position uncertainties for low parallax features are not well represented by the Gaussian distributions implicit in the EKF. The depth coordinate of such features has a probability density which rises sharply at a welldefined minimum depth to a peak, but then tails off very slowly towards infinity –from low parallax measurements it is very difficult to tell whether a feature has a depth of 10 units rather than 100, 1000 or more.

This chapter describes a new feature parametrization which is able to cope smoothly with initialization of features at all depths –even up to "infinity"– within the standard EKF framework. The key concepts are: First, the direct parametrization of inverse depth relative to the camera position from which a feature was first observed. And second, the storage in the state vector of the camera position when the feature was first seen.

3.1.1 Delayed and Undelayed Initialization

The most obvious approach to dealing with feature initialization within a monocular SLAM system is to treat newly detected features separately from the main map, accumulating information in special processing over several frames to reduce depth uncertainty before insertion into the full filter with a standard XYZ representation. Such *delayed initialization* schemes (e.g. [Davison 2003, Kim & Sukkarieh 2003, Bryson & Sukkarieh 2005]) have the drawback that new features, held outside the main probabilistic state, are not able to contribute to the estimation of the camera position until finally included in the map. Further, features which retain low parallax over many frames (those very far from the camera, or close to the motion epipole) are usually rejected completely because they never pass the test for inclusion. In this case, far features cannot be represented in the main SLAM map and their valuable information cannot be incorporated to the filter.

In the delayed approach of Bailey [Bailey 2003], initialization is delayed until the measurement equation is approximately Gaussian and the point can be safely triangulated; here the problem was posed in 2D and validated in simulation. A similar approach for 3D monocular vision with inertial sensing was proposed in [Bryson & Sukkarieh 2005]. Davison [Davison 2003] reacted to the detection of a new feature by inserting a 3D semi-infinite ray into the main map representing everything about the feature except its depth, and then used an auxiliary particle filter to explicitly refine the depth estimate over several frames, taking advantage of all the measurements in a high frame-rate sequence but again with new features held outside the main state vector until inclusion.

More recently, several *undelayed initialization* schemes have been proposed, which still treat new features in a special way but are able to benefit immediately from them to improve camera motion estimates — the key insight being that while features with highly uncertain depths provide little information on camera translation, they are extremely useful as bearing references for orientation estimation. The undelayed method proposed by Kwok and Dissanayake [Kwok & Dissanayake 2004] was a multiple hypothesis scheme, initializing features at various depths and pruning those not reobserved in subsequent images.

[Solà *et al.* 2005, Solà 2007] described a more rigorous undelayed approach using a Gaussian Sum Filter approximated by a Federated Information Sharing method to keep the computational overhead low. An important insight was to spread the Gaussian depth hypotheses along the ray according to inverse depth, achieving much better representational efficiency in this way. This method can perhaps be seen as the direct stepping stone between Davison's particle method and our new inverse depth scheme; a Gaussian sum is a more efficient representation than particles (efficient enough that the separate Gaussians can all be put into the main state vector), but not as efficient as the single Gaussian representation that the inverse depth parametrization allows. Note that neither [Kwok & Dissanayake 2004] nor [Solà *et al.* 2005] consider features at very large 'infinite' depths.

3.1.2 Points at Infinity

The homogeneous coordinate systems of visual projective geometry used normally in SFM allow explicit representation of points at infinity, and they have proven to play an important role during off-line structure and motion estimation. In a sequential SLAM system, the difficulty is that we do not know in advance which features are infinite and which are not. Chapter 2 has shown that in the special case where all features are known to be infinite –in very large scale outdoor scenes or when the camera rotates on a tripod– SLAM in pure angular coordinates turns the camera into a real-time visual compass. In the more general case, let us imagine a camera moving through a 3D scene with observable features at a range of depths. From the estimation point of view, we can think of all features starting at infinity and "coming in" as the camera moves far enough to measure sufficient parallax. For nearby indoor features, only a few centimetres of movement will be sufficient. Distant features may require many meters or even kilometers of motion before parallax is observed. It is important that these features are not permanently labelled as infinite –a feature that seems to be at infinity should always have the chance to prove its finite depth given enough motion, or there will be the serious risk of systematic errors in the scene map. Our probabilistic SLAM algorithm must be able to represent the uncertainty in depth of seemingly infinite features. Observing no parallax for a feature after 10 units of camera translation does tell us something about its depth –it gives a reliable lower bound, which depends on the amount of motion made by the camera (if the feature had been closer than this we would have observed parallax). This explicit consideration of uncertainty in the locations of points has not been previously required in off-line computer vision algorithms, but is very important in the on-line case.

3.1.3 Inverse Depth Representation

Our contribution is to show that in fact there is a unified and straightforward parametrization for feature locations which can handle both initialisation and standard tracking of both close and very distant features within the standard EKF framework. An explicit parametrization of the *inverse depth* of a feature along a semi-infinite ray from the position from which it was first viewed allows a Gaussian distribution to cover uncertainty in depth which spans a depth range from nearby to infinity, and permits seamless crossing over to finite depth estimates of features which have been apparently infinite for long periods of time. The unified representation means that the EKF requires no special initialisation process for features. They are simply tracked right from the start, immediately contribute to improved camera estimates and have their correlations with all other features in the map correctly modelled. Note that the inverse depth parameterization would be equally compatible with other variants of Gaussian filtering such as sparse information filters.

We also introduce in this chapter a linearity index and use it to analyze and prove the representational capability of the inverse depth parametrization for both low and high-parallax features. The only drawback of the inverse depth scheme is the computational issue of increased state vector size, since an inverse depth point needs six parameters rather than the three of XYZ coding. As a solution to this, we show that our linearity index can also be applied to the XYZ parametrization to signal when a feature can be safely switched from inverse depth to XYZ; the usage of the inverse depth representation can in this way be restricted to low parallax feature cases where the XYZ encoding departs from Gaussianity. Note that this 'switching', unlike in delayed initialization methods, is purely to reduce computational load; SLAM accuracy with or without switching is almost the same.

3.1.4 Inverse Depth in Computer Vision and Tracking

Inverse depth is a concept used widely in computer vision: it appears in the relation between image disparity and point depth in stereo vision; it is interpreted as the parallax with respect to the plane at infinity in [Hartley & Zisserman 2004]. Inverse depth is also used to relate the motion field induced by scene points with the camera velocity in optical flow analysis [Heeger & Jepson 1992]. In the tracking community, 'modified polar coordinates' [Aidala & Hammel 1983] also exploit the linearity properties of the inverse depth representation in the slightly different, but closely related, problem of target motion analysis (TMA) from measurements gathered by a bearing-only sensor with known motion.

However, the inverse depth idea has not previously been properly integrated in sequential, probabilistic estimation of motion and structure. It has been used in EKF based sequential depth estimation from camera known motion [Matthies *et al.* 1989] and in multi-baseline stereo Okutomi and Kanade [Okutomi & Kanade 1993] used the inverse depth to increase matching robustness for scene symmetries; matching scores coming from multiple stereo pairs with different baselines were accumulated in a common reference coded in inverse depth, this paper focusing on matching robustness and not on probabilistic uncertainty propagation. In [Chowdhury & Chellappa 2003] it is proposed a sequential EKF process using inverse depth but this was some way short of full SLAM in its details. Images are first processed pairwise to obtain a sequence of 3D motions which are then fused with an individual EKF per feature.

It is our parametrization of inverse depth *relative to the positions from which features were first observed* which means that a Gaussian representation is uniquely well behaved, and this is the reason why a straighforward parametrization of monocular SLAM in the homogeneous coordinates of SFM will not give a good result — that representation only meaningfully represents points which appear to be infinite relative to the coordinate origin. It could be said in projective terms that our method defines separate but correlated projective frames for each feature. Another interesting comparison is between this inverse depth representation, where the representation for each feature includes the camera position from which it was first observed and smoothing/Full SLAM schemes where all historical sensor pose estimates are maintained in a filter.

There are two works that appeared simultaneously with the one presented here – first presented in [Montiel *et al.* 2006, Civera *et al.* 2007a, Civera *et al.* 2008a]– that share the underlying idea of the inverse depth coding. Trawny and Roumeliotis in [Trawny & Roumeliotis 2006] proposed an undelayed initialization for 2D monocular SLAM which encodes a map point as the intersection of two projection rays. This representation is overparametrized but allows undelayed initialization and encoding of both close and distant features, the approach validated with simulation results.

Eade and Drummond presented an inverse depth initialisation scheme within the context of their FastSLAM-based system for monocular SLAM [Eade & Drummond 2006], offering some of the same arguments about advantages in linearity as in this chapter. The position of each new partially initialised feature added to the map is parametrized with three coordinates representing its direction and inverse depth relative to the camera pose at the first observation, and estimates of these coordinates are refined within a set of Kalman Filters for each particle of the map. Once the inverse depth estimation has collapsed, the feature is converted to a fully initialised standard XYZ representation. While retaining the differentiation between partially and fully-initialised features, they go further and are able to use measurements of partially initialised features with unknown depth to improve estimates of camera orientation and translation via a special epipolar update step. Their approach certainly appears appropriate within a FastSLAM implementation. However, it lacks the satisfying unified quality of the parametrization we present in this chapter, where the transition from partially to fully initialised need not be explicitly tackled and full use is automatically made of all of the information available in measurements.

The rest of the chapter is organized as follows: Section 3.2 defines the state vector, including the camera motion model, XYZ point coding and inverse depth point parametrization. The measurement equation is described in section 3.3. Section 3.4 presents the linearity index we will use for further insight on the superiority of the inverse depth coding and for conversion to XYZ parametrization. Feature initialization from a single feature observation is detailed in section 3.5. In section 3.6 the switch from inverse depth to XYZ coding is presented. Section 3.7 details of the explotation of the Bayesian priors in the matching step: the patch warping algorithm in subsection 3.7.1 and the active search for correspondences [Davison *et al.* 2007] in subsection 3.7.2. In section 3.8 experimental validation over real image sequences captured at 30Hz in large scale environments, indoors and outdoors, including real-time performance and a loop closing experiment is presented. Finally, section 3.9 is devoted to summarize and discuss the contributions in the chapter.

3.2 State Vector Definition

3.2.1 Camera Motion

A constant angular and linear velocity model is used to model hand-held camera motion. The camera state $\mathbf{x}_C$ is composed of pose terms: $\mathbf{r}^{WC}$ camera optical center position and $\mathbf{q}^{WC}$ quaternion defining orientation; and linear and angular velocity $\mathbf{v}^W$ and ω^C relative to world frame W and camera frame C.

We assume that linear and angular accelerations $\mathbf{a}^W$ and α^C affect the camera, producing at each step an impulse of linear velocity, $\mathbf{V}^W = \mathbf{a}^W \Delta t$, and angular velocity $\Omega^C = \alpha^C \Delta t$, with zero mean and known Gaussian distribution. We currently assume a diagonal covariance matrix for the unknown input linear and angular accelerations.

The state update equation for the camera is:

$$\mathbf{x}_{C_{k+1}} = \begin{pmatrix} \mathbf{r}^{WC}_{k+1} \\ \mathbf{q}^{WC}_{k+1} \\ \mathbf{v}^{W}_{k+1} \\ \omega^{C}_{k+1} \end{pmatrix} = \mathbf{f}_v\left(\mathbf{x}_{C_k}, \mathbf{n}\right) = \begin{pmatrix} \mathbf{r}^{WC}_{k} + \left(\mathbf{v}^{W}_{k} + \mathbf{V}^{W}_{k}\right)\Delta t \\ \mathbf{q}^{WC}_{k} \times \mathbf{q}\left(\left(\omega^{C}_{k} + \Omega^{C}\right)\Delta t\right) \\ \mathbf{v}^{W}_{k} + \mathbf{V}^{W} \\ \omega^{C}_{k} + \Omega^{C} \end{pmatrix} . \tag{3.1}$$

where $\mathbf{q}\left(\left(\omega^{C}_{k} + \Omega^{C}\right)\Delta t\right)$ is the quaternion defined by the rotation vector $\left(\omega^{C}_{k} + \Omega^{C}\right)\Delta t$.

3.2.2 Euclidean XYZ Point Parametrization

The standard representation for scene points i in terms of Euclidean XYZ coordinates (see Figure 3.1) is:

$$\mathbf{y}^{W}_{\mathrm{XYZ},i} = \left(X_i\ Y_i\ Z_i\right)^\top . \tag{3.2}$$

Along the book, we sometimes refer to the Euclidean XYZ coding simply as XYZ coding.

3.2.3 Inverse Depth Point Parametrization

In the model presented in this chapter, a 3D point in the scene i is defined by a state vector with 6 parameters:

$$\mathbf{y}^{W}_{\rho,i} = \left(x_i\ y_i\ z_i\ \theta_i\ \phi_i\ \rho_i\right)^\top , \tag{3.3}$$

which Cartesian coordinates are (see Figure 3.1):

$$\begin{pmatrix} X_i \\ Y_i \\ Z_i \end{pmatrix} = \begin{pmatrix} x_i \\ y_i \\ z_i \end{pmatrix} + \frac{1}{\rho_i}\mathbf{m}\left(\theta_i, \phi_i\right) \tag{3.4}$$

$$\mathbf{m} = \begin{pmatrix} \cos\phi_i \sin\theta_i & -\sin\phi_i & \cos\phi_i \cos\theta_i \end{pmatrix}^\top . \tag{3.5}$$

The vector $\mathbf{y}_{\rho,i}^{W}$ encodes the *ray from the first camera position from which the feature was observed* by x_i, y_i, z_i, the camera optical center, and θ_i, ϕ_i azimuth and elevation (coded in the world frame) defining unit directional vector $\mathbf{m}\left(\theta_i, \phi_i\right)$. The point's depth along the ray d_i is encoded by its inverse $\rho_i = 1/d_i$.

It is important to remark here that this chapter represents every geometric entity with respect to a static world reference frame W. Referring every camera and feature parameters to a dynamic reference frame C_k attached to the camera improves results when using the EKF [Castellanos *et al.* 2004]; so next chapter will also consider this other case.

3.2.4 Full State Vector

As in standard EKF SLAM, we use a single joint state vector containing camera pose and feature estimates, with the assumption that the camera moves with respect to a static scene. The whole state vector $\mathbf{x}$ is composed of the camera and all the map features:

$$\mathbf{x} = \left(\mathbf{x}_C^\top, \mathbf{y}_1^\top, \ldots \mathbf{y}_i^\top, \ldots \mathbf{y}_n^\top\right)^\top . \tag{3.6}$$

3.3 Measurement Equation

Each observed feature imposes a constraint between the camera location and the corresponding map feature (see Figure 3.1). Observation of a point $\mathbf{y}_i$ ($\mathbf{x}_i$) defines a ray coded by a directional vector in the camera frame $\mathbf{h}^C = \begin{pmatrix} h_x & h_y & h_z \end{pmatrix}^\top$. For points in XYZ:

$$\mathbf{h}^C = \mathbf{h}_{\mathrm{XYZ}}^C = \mathbf{R}^{CW}\left(\begin{pmatrix} X_i \\ Y_i \\ Z_i \end{pmatrix} - \mathbf{r}^{WC}\right) . \tag{3.7}$$

For points in inverse depth:

$$\mathbf{h}^C = \mathbf{h}_\rho^C = \mathbf{R}^{CW}\left(\rho_i\left(\begin{pmatrix} x_i \\ y_i \\ z_i \end{pmatrix} - \mathbf{r}^{WC}\right) + \mathbf{m}\left(\theta_i, \phi_i\right)\right) , \tag{3.8}$$

where the directional vector has been normalized using the inverse depth. It is worth noting that (3.8) can be safely used even for points at infinity i.e $\rho_i = 0$.

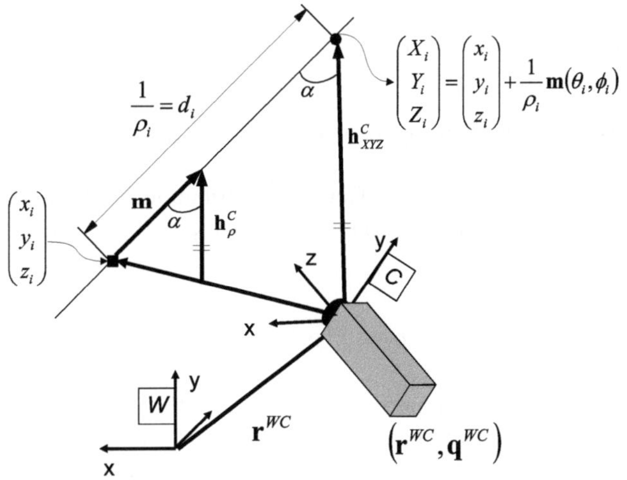

Fig. 3.1. Feature parametrization and measurement equation.

The camera does not directly observe $\mathbf{h}^C$ but its projection in the image according to the pinhole model. Projection to a normalized retina and then camera calibration is applied:

$$\mathbf{h} = \begin{pmatrix} u \\ v \end{pmatrix} = \begin{pmatrix} C_x - \frac{f}{d_x}\frac{h_x}{h_z} \\ C_y - \frac{f}{d_y}\frac{h_y}{h_z} \end{pmatrix}, \tag{3.9}$$

where $(C_x \; C_y)^\top$ is the camera's principal point, f is the focal length and $(d_x \; d_y)^\top$ the pixel size. Finally, a distortion model has to be applied to deal with real camera lenses. In this work we have used the standard two parameter distortion model from photogrammetry [Mikhail *et al.* 2001] (see Appendix for details.)

It is worth noting that the measurement equation in inverse depth has a sensitive dependency on the parallax angle α (see Figure 3.1). At low parallax, Equation (3.8) can be approximated by $\mathbf{h}^C \approx \mathbf{R}^{CW}(\mathbf{m}(\theta_i, \phi_i))$, and hence the measurement equation only provides information about the camera orientation and the directional vector $\mathbf{m}(\theta_i, \phi_i)$.

3.4 Measurement Equation Linearity

The higher the degree of linearity of the measurement equation is, the better the Kalman Filter performs. This section presents an analysis of the degree of linearity for both XYZ and inverse depth parametrizations. These linearity analyses theoretically support the superiority of the inverse depth coding.

3.4.1 Linearized Propagation of a Gaussian

Let x be an uncertain variable with Gaussian distribution $x \sim N\left(\mu_x, \sigma_x^2\right)$. The transformation of x through the function f is a variable y which can be approximated with a Gaussian distribution

$$y \sim N\left(\mu_y, \sigma_y^2\right), \; \mu_y = f\left(\mu_x\right), \; \sigma_y^2 = \left.\frac{\partial f}{\partial x}\right|_{\mu_x} \sigma_x^2 \left.\frac{\partial f}{\partial x}\right|_{\mu_x}^{\top}, \tag{3.10}$$

if the function f is linear in an interval around μ_x (Figure 3.2). The interval size in which the function has to be linear depends on σ_x; the bigger σ_x the wider the interval has to be to cover a significant fraction of the random variable x values. In this work we fix the linearity interval to the 95% confidence region defined by $[\mu_x - 2\sigma_x, \mu_x + 2\sigma_x]$.

If a function is linear in an interval, the first derivative is constant in that interval. To analyze the first derivative variation around the interval $[\mu_x - 2\sigma_x, \mu_x + 2\sigma_x]$ consider the Taylor expansion for the *first derivative*:

$$\frac{\partial f}{\partial x}\left(\mu_x + \Delta x\right) \approx \left.\frac{\partial f}{\partial x}\right|_{\mu_x} + \left.\frac{\partial^2 f}{\partial x^2}\right|_{\mu_x} \Delta x \,. \tag{3.11}$$

We propose to compare the value of the derivative at the interval center, μ_x, with the value at the extremes $\mu_x \pm 2\sigma_x$, where the deviation from linearity will be maximal, using the following dimensionless linearity index:

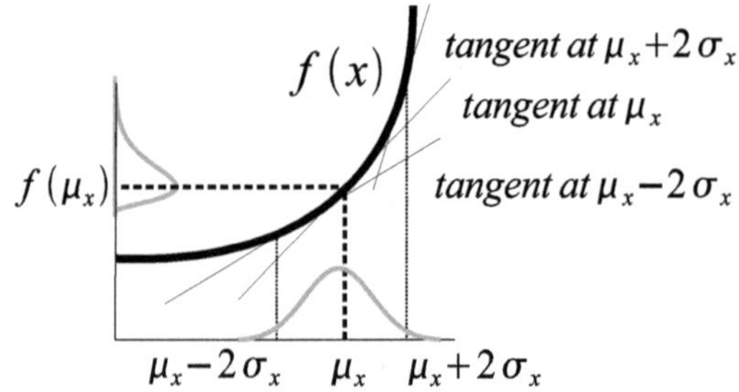

Fig. 3.2. The first derivative variation in $[\mu_x - 2\sigma_x, \mu_x + 2\sigma_x]$ codes the departure from Gaussianity in the propagation of the uncertain variable through a function.

$$L = \left| \frac{\left.\frac{\partial^2 f}{\partial x^2}\right|_{\mu_x} 2\sigma_x}{\left.\frac{\partial f}{\partial x}\right|_{\mu_x}} \right| . \tag{3.12}$$

When $L \approx 0$, the function can be considered linear in the interval, and hence Gaussianity is preserved during transformation.

3.4.2 Linearity of XYZ Parametrization

The linearity of the XYZ representation is analyzed by means of a simplified model which only estimates the depth of a point with respect to the camera. In our analysis, a scene point is observed by two cameras (Figure 3.3a), both of which are oriented towards the point. The first camera detects the ray on which the point lies. The second camera observes the same point from a distance d_1; the parallax angle α is approximated by the angle between the cameras' optic axes.

The point's location error, d, is encoded as Gaussian in depth:

$$D = d_0 + d, \quad d \sim N\left(0, \sigma_d^2\right) . \tag{3.13}$$

This error d is propagated to the image of the point in the second camera, u as:

$$u = \frac{x}{y} = \frac{d \sin\alpha}{d_1 + d\cos\alpha} . \tag{3.14}$$

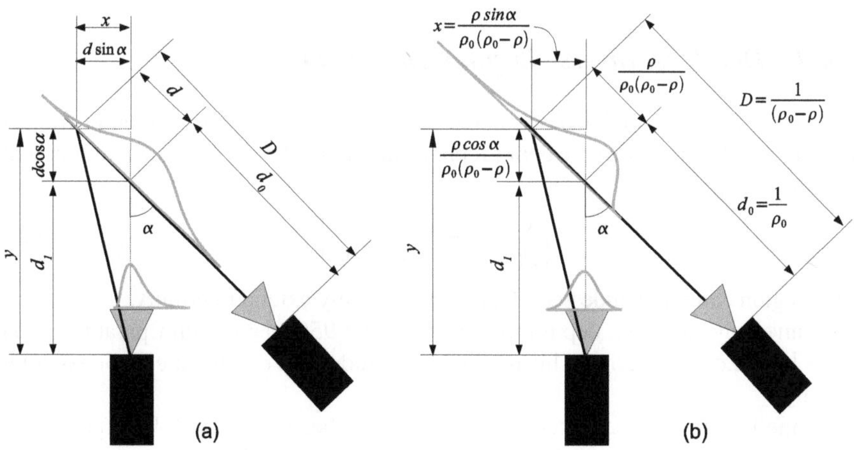

Fig. 3.3. Uncertainty propagation from the scene point to the image. (a) XYZ coding. (b) Inverse depth coding.

The Gaussianity of u is analyzed by means of (3.12), giving linearity index:

$$L_d = \left| \frac{\frac{\partial^2 u}{\partial d^2} 2\sigma_d}{\frac{\partial u}{\partial d}} \right| = \frac{4\sigma_d}{d_1} \left| cos\alpha \right| \tag{3.15}$$

3.4.3 Linearity of Inverse Depth Parametrization

The inverse depth parametrization is based on the same scene geometry as the direct depth coding, but the depth error is encoded as Gaussian in inverse depth (Figure 3.3b):

$$D = \frac{1}{\rho_0 - \rho}, \quad \rho \sim N\left(0, \sigma_\rho^2\right) \tag{3.16}$$

$$d = D - d_0 = \frac{\rho}{\rho_0 \left(\rho_0 - \rho\right)}, \quad d_0 = \frac{1}{\rho_0} . \tag{3.17}$$

So the image of the scene point is computed as:

$$u = \frac{x}{y} = \frac{d \sin\alpha}{d_1 + d\cos\alpha} = \frac{\rho \sin\alpha}{\rho_0 d_1 \left(\rho_0 - \rho\right) + \rho \cos\alpha}, \tag{3.18}$$

and the linearity index L_ρ is now:

$$L_\rho = \left| \frac{\frac{\partial^2 u}{\partial \rho^2} 2\sigma_\rho}{\frac{\partial u}{\partial \rho}} \right| = \frac{4\sigma_\rho}{\rho_0} \left| 1 - \frac{d_0}{d_1} cos\alpha \right| . \tag{3.19}$$

3.4.4 Depth vs. Inverse Depth Comparison

When a feature is initialized, the depth prior has to cover a vast region in front of the camera. With the inverse depth representation, the 95% confidence region with parameters ρ_0, σ_ρ is:

$$\left[\frac{1}{\rho_0 + 2\sigma_\rho}, \frac{1}{\rho_0 - 2\sigma_\rho} \right] . \tag{3.20}$$

This region cannot include zero depth but can easily extend to infinity.

Conversely, with the depth representation the 95% region with parameters d_0, σ_d is $\left[d_0 - 2\sigma_d, d_0 + 2\sigma_d\right]$. This region can include zero depth but cannot extend to infinity.

In the first few frames after a new feature has been initialized, little parallax is likely to have been observed. Therefore $\frac{d_0}{d_1} \approx 1$ and $\alpha \approx 0 \Longrightarrow \cos\alpha \approx 1$. In this case the L_d linearity index for depth is high (bad), while the L_ρ linearity index for inverse depth is low (good): during initialization the inverse depth measurement equation linearity is superior to the XYZ coding.

As estimation proceeds and α increases, leading to more accurate depth estimates, the inverse depth representation continues to have a high degree of linearity. This is because in the expression for L_ρ the increase in the term $\left|1-\frac{d_0}{d_1}cos\alpha\right|$ is compensated by the decrease in $\frac{4\sigma_\rho}{\rho_0}$. For inverse depth features a good linearity index is achieved along the whole estimation history. So the inverse depth coding is suitable for both low and high parallax cases if the feature is continuously observed.

The XYZ encoding has low computational cost, but achieves linearity only at low depth uncertainty and high parallax. In section 3.6 we explain how the representation of a feature can be switched over such that the inverse depth parametrization is only used when needed — for features which are either just initialized or at extreme depths.

3.5 Feature Initialization

From just a single observation no feature depth can be estimated (although it would be possible in principle to impose a very weak depth prior by knowledge of the type of scene observed). What we do is to assign a general Gaussian prior in inverse depth which encodes probabilistically the fact that the point has to be in front of the camera. Hence, thanks to the linearity of inverse depth at low parallax, the filter can be initialized from just one observation. Experimental tuning has shown that infinity should be included with reasonable probability within the initialization prior, despite the fact that this means that depth estimates can become negative. Once initialized, features are processed with the standard EKF prediction-update loop — even in the case of negative inverse depth estimates — and immediately contribute to camera location estimation within SLAM.

It is worth noting that while a feature retains low parallax, it will automatically be used mainly to determine the camera orientation. The feature's depth will remain uncertain, with the hypothesis of infinity still under consideration (represented by the probability mass corresponding to negative inverse depths). If the camera translates to produce enough parallax then the feature's depth estimation will be improved and it will begin to contribute more to camera location estimation.

The initial location for a newly observed feature inserted into the state vector is:

$$\hat{\mathbf{y}}\left(\hat{\mathbf{r}}^{WC},\hat{\mathbf{q}}^{WC},\mathbf{z}_i,\rho_0\right)=\left(\hat{x}_i\ \hat{y}_i\ \hat{z}_i\ \hat{\theta}_i\ \hat{\phi}_i\ \hat{\rho}_i\right)^\top , \tag{3.21}$$

a function of the current camera pose estimate $\hat{\mathbf{r}}^{WC},\hat{\mathbf{q}}^{WC}$, the image observation $\mathbf{z}_i=\left(u_i\ v_i\right)^\top$ and the parameters determining the depth prior ρ_0, σ_{ρ_0}.

The end-point of the initialization ray (see Figure 3.1) is taken from the current camera location estimate:

$$\left(\hat{x}_i \; \hat{y}_i \; \hat{z}_i \right)^\top = \hat{\mathbf{r}}^{WC} , \tag{3.22}$$

and the direction of the ray is computed from the observed point, expressed in the world coordinate frame:

$$\mathbf{h}^W = \mathbf{R}_{WC} \left(\hat{\mathbf{q}}^{WC} \right) \left(\upsilon_i \; \nu_i \; 1 \right)^\top , \tag{3.23}$$

where υ_i and ν_i are normalized retina image coordinates. Despite $\mathbf{h}^W$ being a non-unit directional vector, the angles by which we parametrize its direction can be calculated as:

$$\begin{pmatrix} \hat{\theta}_i \\ \hat{\phi}_i \end{pmatrix} = \begin{pmatrix} \arctan \left(h_x^W , h_z^W \right) \\ \arctan \left(-h_y^W , \sqrt{h_x^{W\,2} + h_z^{W\,2}} \right) \end{pmatrix} . \tag{3.24}$$

The covariance of $\hat{x}_i, \hat{y}_i, \hat{z}_i$, $\hat{\theta}_i$ and $\hat{\phi}_i$ is derived from the image measurement error covariance $\mathbf{R}_i$ and the state covariance estimate $\hat{\mathbf{P}}_{k|k}$.

The initial value for ρ_0 and its standard deviation are set such that the 95% confidence region spans a range of depths from a close distance to the camera up to infinity. In our experiments of this chapter we chose $\hat{\rho}_0 = 0.1, \sigma_{\rho_0} = 0.5$, which gives an inverse depth confidence region $[1.1, -0.9]$. Notice that infinity is included in this range. Nevertheless, further experimental validation has shown that the precise values of these parameters are relatively unimportant to the accurate operation of the filter as long as infinity is clearly included in the confidence interval.

The state covariance after feature initialization is:

$$\mathbf{P}_{k|k}^{\text{new}} = \mathbf{J} \begin{pmatrix} \mathbf{P}_{k|k} & 0 & 0 \\ 0 & \mathbf{R}_i & 0 \\ 0 & 0 & \sigma_{\rho_0}^2 \end{pmatrix} \mathbf{J}^\top \tag{3.25}$$

$$\mathbf{J} = \left(\begin{array}{c|c} I & 0 \\ \hline \frac{\partial \mathbf{y}}{\partial \mathbf{r}^{WC}}, \frac{\partial \mathbf{y}}{\partial \mathbf{q}^{WC}}, 0, \ldots, 0, & \frac{\partial \mathbf{y}}{\partial \mathbf{z}_i}, \frac{\partial \mathbf{y}}{\partial \rho} \end{array} \right) . \tag{3.26}$$

The inherent scale ambiguity in monocular SLAM has usually been fixed by observing some known initial features that fix the scale (e.g. [Davison 2003]). A very interesting experimental observation we have made using the inverse depth scheme is that sequential monocular SLAM can operate successfully without *any* known features in the scene, and in fact the experiments we present in this book do not use an initialization target. In this case the overall scale of the reconstruction and camera motion is undetermined, although with the formulation of the current chapter the estimation will settle on a (meaningless) scale of some value. Appendix B details an alternative formulation of EKF SfM that illustrates this issue via a dimensional analysis of the problem and the proposal of a dimensionless formulation.

3.6 Switching from Inverse Depth to XYZ

While the inverse depth encoding can be used at both low and high parallax, it is advantageous for reasons of computational efficiency to restrict inverse depth to cases where the XYZ encoding exhibits non linearity according to the L_d index. This section details switching from inverse depth to XYZ for high parallax features.

3.6.1 Conversion from Inverse Depth to XYZ Coding

After each estimation step, the linearity index L_d (Equation 3.15) is computed for every map feature coded in inverse depth:

$$\mathbf{h}^W_{\text{XYZ}} = \hat{\mathbf{y}}^W_{\text{XYZ},i} - \hat{\mathbf{r}}^{WC}, \quad \sigma_d = \frac{\sigma_\rho}{\rho_i^2}, \quad \sigma_\rho = \sqrt{\mathbf{P}_{\mathbf{y}_i\mathbf{y}_i}(6,6)}$$
$$d_i = \left\|\mathbf{h}^W_{\text{XYZ}}\right\|, \quad \cos\alpha = \mathbf{m}^\top \mathbf{h}^W_{\text{XYZ}} \left\|\mathbf{h}^W_{\text{XYZ}}\right\|^{-1}. \tag{3.27}$$

where $\hat{\mathbf{y}}^W_{\text{XYZ},i}$ is computed using equation (3.4) and $\mathbf{P}_{\mathbf{y}_i\mathbf{y}_i}$ is the submatrix 6×6 covariance matrix corresponding the considered feature.

If L_d is below a switching threshold, the feature is transformed using Equation (3.4) and the *full state* covariance matrix $\mathbf{P}$ is transformed with the corresponding Jacobian:

$$\mathbf{P}_{\text{new}} = \mathbf{J}\mathbf{P}\mathbf{J}^\top, \quad \mathbf{J} = \text{diag}\left(\mathbf{I}, \frac{\partial \mathbf{y}_{\text{XYZ},i}}{\partial \mathbf{y}_{\rho,i}}, \mathbf{I}\right). \tag{3.28}$$

3.6.2 Linearity Index Threshold

We propose to use index L_d (3.15) to define a threshold for switching from inverse depth to XYZ encoding at the point when the latter can be considered linear. If the XYZ representation is linear, then the measurement u is Gaussian distributed (Equation 3.10):

$$u \sim N\left(\mu_u, \sigma_u^2\right), \quad \mu_u = 0, \quad \sigma_u^2 = \left(\frac{\sin\alpha}{d_1}\right)^2 \sigma_d^2. \tag{3.29}$$

To determine the threshold in L_d which signals a lack of linearity in the measurement equation a simulation experiment has been performed. The goal was to generate samples from the uncertain distribution for variable u and then apply a standard Kolmogorov-Smirnov Gaussianity test [Canavos 1984] to these samples, counting the percentage of rejected hypotheses h. When u is effectively Gaussian, the percentage should match the test significance level α_{sl} (5% in our experiments);

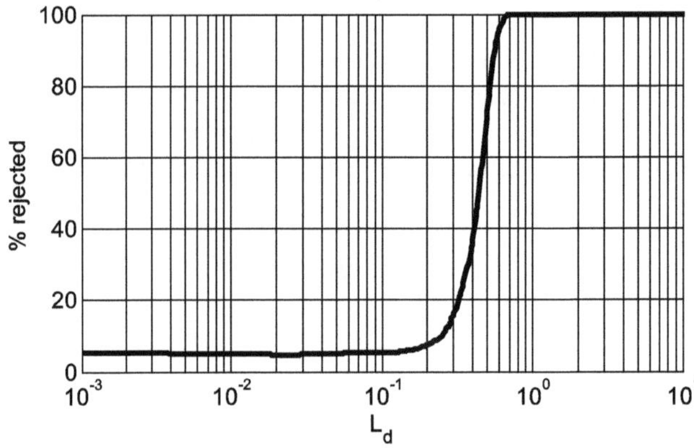

Fig. 3.4. Percentage of test rejections as a function of the linearity index L_d

as the number of rejected hypotheses increases the measurement equation departs from linearity. A plot of the percentage of rejected hypotheses h with respect to the linearity index L_d is shown in Figure 3.4. It can be clearly seen than when $L_d > 0.2$, h sharply departs from 5%. So we propose the $L_d < 10\%$ safe threshold for switching from inverse depth to XYZ encoding.

Notice that the plot in Figure 3.4 is smooth (log scale in L_d), which indicates that the linearity index effectively represents the departure from linearity.

The simulation has been performed for a variety of values of α, d_1 and σ_d; more precisely all triplets resulting from the following parameter values:

$$\alpha(\text{deg}) \in \{0.1, 1, 3, 5, 7, 10, 20, 30, 40, 50, 60, 70\}$$
$$d_1(\text{m}) \in \{1, 3, 5, 7, 10, 20, 50, 100\}$$
$$\sigma_d(\text{m}) \in \{0.05, 0.1, 0.25, 0.5, 0.75, 1, 2, 5\}\ .$$

The simulation algorithm detailed in Figure 3.5 is applied to every triplet $\{\alpha, d_1, \sigma_d\}$ to count the percentage of rejected hypotheses h and the corresponding linearity index L_d.

3.7 Data Association

The information contained in the updated probability distribution that filtering-based approaches maintain along the image sequence can be exploited in the matching step, resulting in an increase of robustness at a lower computational cost. Following the terminology in [Williams *et al.* 2008], correspondences are not searched in an image–to–image but in an image–to–map basis that takes into account the prior knowledge over the structure of the scene and predicted camera motion.

input: α, d_1, σ_d
output: h, L_d
$\sigma_u = \left|\frac{\sin\alpha}{d_1}\right| \sigma_d$; $\mu_u = 0$; //(3.29)
$\alpha_{sl} = 0.05$; // Kolm. test sign. level
$L_d = \frac{4\sigma_d}{d_1} |\cos\alpha|$
n_rejected=0 ;
N_GENERATED_SAMPLES=1000;
SAMPLE_SIZE=1000;

for j=1 **to** N_GENERATED_SAMPLES **repeat**
  $\{d_i\}_j$=random_normal(0, σ_d^2, SAMPLE_SIZE);
    //generate a normal sample from $N(0,\sigma_d^2)$;
  $\{u_i\}_j$=propagate_from_dept_to_image($\{d_i\}_j$, α, d_1);//(3.14)
  if rejected==Kolmogorov_Smirnov($\{u_i\}_j, \mu_u, \sigma_u, \alpha_{sl}$)
    n_rejected=n_rejected+1;
endfor
$h = 100 \frac{\lceil n\_rejected \rceil}{\lceil N\_GENERATED\_SAMPLES \rceil}$;

Fig. 3.5. Simulation algorithm to test the linearity of the measurement equation.

Using this prior information, the patch which serves as photometric identifier of the feature can be warped according to the predicted camera motion facilitating the correspondence search under large projective distortions. Also, the propagation of the probabilistic density over the state through the projection model can be used to define small regions of high probability for finding the match –this is called active search in [Davison *et al.* 2007].

3.7.1 Patch Warping

The performance of any SfM system relies on its ability to find correct correspondences of a 3D feature in several images. Finding such correspondences succesfully becomes more difficult as the motion between images increases, as the appearance of the feature in the image strongly vary. Pairwise SfM methods, that may take as input widely separated images, usually make use of descriptors with a high degree of invariance –at the cost of a more expensive processing– [Lowe 2004, Bay *et al.* 2008, Mikolajczyk *et al.* 2005].

The sequential processing of video sequences of this book makes the use of these invariant descriptors unnecesary. Although sequential SfM from videos also faces the problem of searching correspondences in widely separated images, it can benefit from the registration of past images and a dynamic motion between frames. As a difference from pairwise SfM, the correspondence search does not have to start "from scratch" without knowing the scale or rotation for each match. The appearance of each feature can be accurately predicted using the relative motion between cameras and the 3D feature.

The use of highly invariant descriptors does not offer any advantages in sequential SfM or monocular SLAM; and it is advisable to use simple descriptors and predict their projective distortion using the geometric estimation. In [Chekhlov *et al.* 2007], patches at different scales are generated when a feature is first detected and are selected in the matching stage according to the predicted scale change. In [Molton *et al.* 2004], the patch is warped according to the predicted motion and also the normal to the local patch is estimated.

In the experiments of this book, the feature patches are warped according to the estimation in order to improve the matching. Differently from [Molton *et al.* 2004], we do not estimate the normal of the patch but approximate it by the bisector of the angle formed by the two projection rays. We find that this simplification successfully copes with the most problematic motions (e.g., cyclotorsion) and its general performance do not degrade much compared with more elaborated approaches. The warping or the patch is computed via the following homography

$$\mathbf{H} = \mathbf{K}\left(\mathbf{R}^{C_iC_k} - \mathbf{t}^{C_iC_k}\mathbf{n}^\top / d_i\right)\mathbf{K}^{-1}\,, \tag{3.30}$$

where $\mathbf{K}$ stands for the known calibration matrix.

$$\mathbf{K} = \begin{pmatrix} f/d_x & 0 & C_x \\ 0 & f/d_y & C_y \\ 0 & 0 & 1 \end{pmatrix}\,. \tag{3.31}$$

$\mathbf{R}^{C_iC_k}$ and $\mathbf{t}^{C_iC_k}$ are the relative rotation and translation between the camera C_i where the feature was initialised and the current one C_k. Both can be computed from the current camera rotation $\mathbf{R}^{WC_k}$ and translation $\mathbf{t}^{WC_k}$ and the ones at the feature initialization $\mathbf{R}^{WC_i}$ and $\mathbf{t}^{WC_i}$. d_i is the distance from the first camera to the point, that can be extracted easily from the state vector –it is the inverse of the inverse depth for this feature ρ_i. $\mathbf{n}$ stands for the normal of the patch, and it is approximated as described in the previous paragraph.

3.7.2 Active Search

Active search or guided matching is achieved in filtering-based visual estimation by projecting into the image the probabilistic state vector. The 95% probability region from this distribution results in small elliptical search regions. Searching the correspondences inside the ellipses reduces the computational cost and also prevents many outliers to appear. Although the spurious rate reduces drastically by this approach, it is important to remark that a complete rejection of the outliers cannot be guaranteed and an additional step that checks the consensus of the data against a global model is still needed. Next chapter further elaborates on this issue and presents an efficient matching algorithm based on RANSAC.

The search region is obtained by propagating first the EKF Gaussian prediction $\mathcal{N}\left(\hat{\mathbf{x}}_{k|k-1}, \mathbf{P}_{k|k-1}\right)$ through the measurement equation $\mathbf{h}(\mathbf{x})$, resulting in a multidimensional Gaussian in the image $\mathcal{N}\left(\hat{\mathbf{h}}_{k|k-1}, \mathbf{S}_{k|k-1}\right)$

$$\hat{\mathbf{h}}_{k|k-1} = \mathbf{h}(\hat{\mathbf{x}}_{k|k-1}) \tag{3.32}$$

$$\mathbf{S}_{k|k-1} = \mathbf{H}_{k|k-1}\mathbf{P}_{k|k-1}\mathbf{H}_{k|k-1}^{\top} + \mathbf{R}_k \,. \tag{3.33}$$

$\mathbf{H}_{k|k-1}$ are the derivatives for the measurement equation in the prediction and $\mathbf{R}_k$ the image noise covariance at step k. The individual search regions for a feature i are the parameters corresponding to the 2D mean $\hat{\mathbf{h}}_i$ and covariance $\mathbf{S}_i$ extracted from the multivariate distribution.

It can be seen in any of the figures in the experimental results section where images are displayed –Figures 3.6, 3.8 and 3.10– and other figures along the book –e.g. Figures 2.9, 4.10 and 6.4– the small ellipses around feature image predictions where the matches are looked for. Efficiency gain come across straightforwardly if it is compared the size of these ellipses with the size of the whole image.

3.8 Experimental Results

The performance of the inverse depth parametrization has been tested on real image sequences acquired with a hand-held low cost Unibrain IEEE1394 camera, with a 90° field of view and 320×240 resolution, capturing monochrome image sequences at 30 fps.

Five experiments has been performed. The first is an indoor sequence processed offline with a Matlab implementation, the goal being to analyze initialization of scene features located at different depths. The second experiment shows an outdoor sequence processed in real-time with a C++ implementation. The focus here is on

Fig. 3.6. First (a) and last (b) frame in the sequence of the indoor experiment of section 3.8.1. Features 11,12, 13 are analyzed. These features are initialized in the same frame but are located at different distances from the camera.

distant features, observed under low parallax along the whole sequence. The third experiment is a loop closing sequence, concentrating on camera covariance evolution. Fourth is a simulation experiment to analyze the effect of switching from inverse depth to XYZ representations. In the last experiment the switching performance is verified on the real loop closing sequence. The section ends with a computing time analysis.

3.8.1 Indoor Sequence

This experiment analyzes the performance of the inverse depth scheme as several features at a range of depths are tracked within SLAM. We discuss three features, which are all detected in the same frame but have very different depths. Figure 3.6 shows the image where the analyzed features are initialized (frame 18 in the

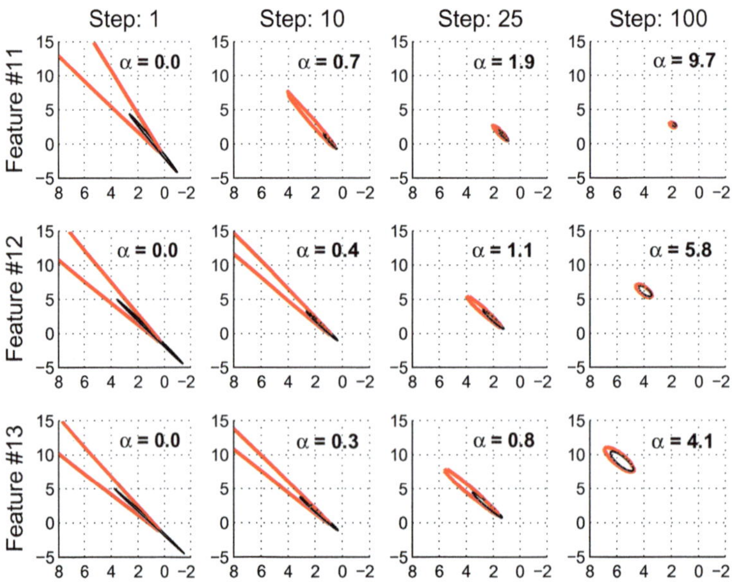

Fig. 3.7. Feature initialization. Each column shows the estimation history for a feature horizontal components. For each feature, the estimates after 1, 10, 25 and 100 frames since initialization are plotted; the parallax angle α in degrees between the initial observation and the current frame is displayed. The thick (red) lines show (calculated by a Monte Carlo numerical simulation) the 95% confidence region when coded as Gaussian in inverse depth. The thin (black) ellipsoids show the uncertainty as a Gaussian in XYZ space propagated according to Equation (3.28). Notice how at low parallax the inverse depth confidence region is very different from the elliptical Gaussian. However, as the parallax increases, the uncertainty reduces and collapses to the Gaussian ellipse.

sequence) and the last image in the sequence. Figure 3.7 focuses on the evolution of the estimates corresponding to the features, with labels 11, 12 and 13, at frames 1, 10, 25 and 100. Confidence regions derived from the inverse depth representation (thick red line) are plotted in XYZ space by numerical Monte Carlo propagation from the six-dimensional multivariate Gaussians representing these features in the SLAM EKF. For comparison, standard Gaussian XYZ acceptance ellipsoids (thin black line) are linearly propagated from the six-dimensional representation by means of the Jacobian of equation (3.28). The parallax α in degrees for each feature at every step is also displayed.

When initialized, the 95% acceptance region of all the features includes $\rho = 0$ so infinite depth is considered as a possibility. The corresponding confidence region in depth is highly asymmetric, excluding low depths but extending to infinity. It is clear that Gaussianity in inverse depth is not mapped to Gaussianity in XYZ, so the black ellipsoids produced by Jacobian transformation are far from representing the true depth uncertainty. As stated in section 3.4.4, it is at low parallax that the inverse depth parametrization plays a key role.

As rays producing bigger parallax are gathered, the uncertainty in ρ becomes smaller but still maps to a non-Gaussian distribution in XYZ. Eventually, at high parallax, for all of the features the red confidence regions become closely Gaussian and well-approximated by the linearly-propagated black ellipses — but this happens much sooner for nearby feature 11 than distant feature 13.

3.8.2 Real-Time Outdoor Sequence

This 860 frame experiment was performed with a C++ implementation which achieves real-time performance at 30 fps with hand-held camera. Here we highlight the ability of our parametrization to deal with both close and distant features in an outdoor setting.

Figure 3.8 shows two frames of the sequence along with the estimation results at those steps. For most of the features, the camera ended up gathering enough parallax to accurately estimate their depths. However, being outdoors, there were distant features producing low parallax during the whole camera motion.

The inverse depth estimation history for two features is highlighted in Figure 3.9. It is shown that distant, low parallax features are persistently tracked through the sequence, despite the fact that their depths cannot be precisely estimated. The large depth uncertainty, represented with the inverse depth scheme, is successfully managed by the EKF SLAM, allowing the orientation information supplied by these features to be exploited.

Feature 3, on a nearby car, eventually gathers enough parallax to have an accurate depth estimate after 250 images where infinite depth was still considered as a possibility. Meanwhile the estimate of feature 11, on a distant tree and never displaying significant parallax, never collapses in this way and zero inverse depth remains within its confidence region. Delayed intialization schemes would have discarded

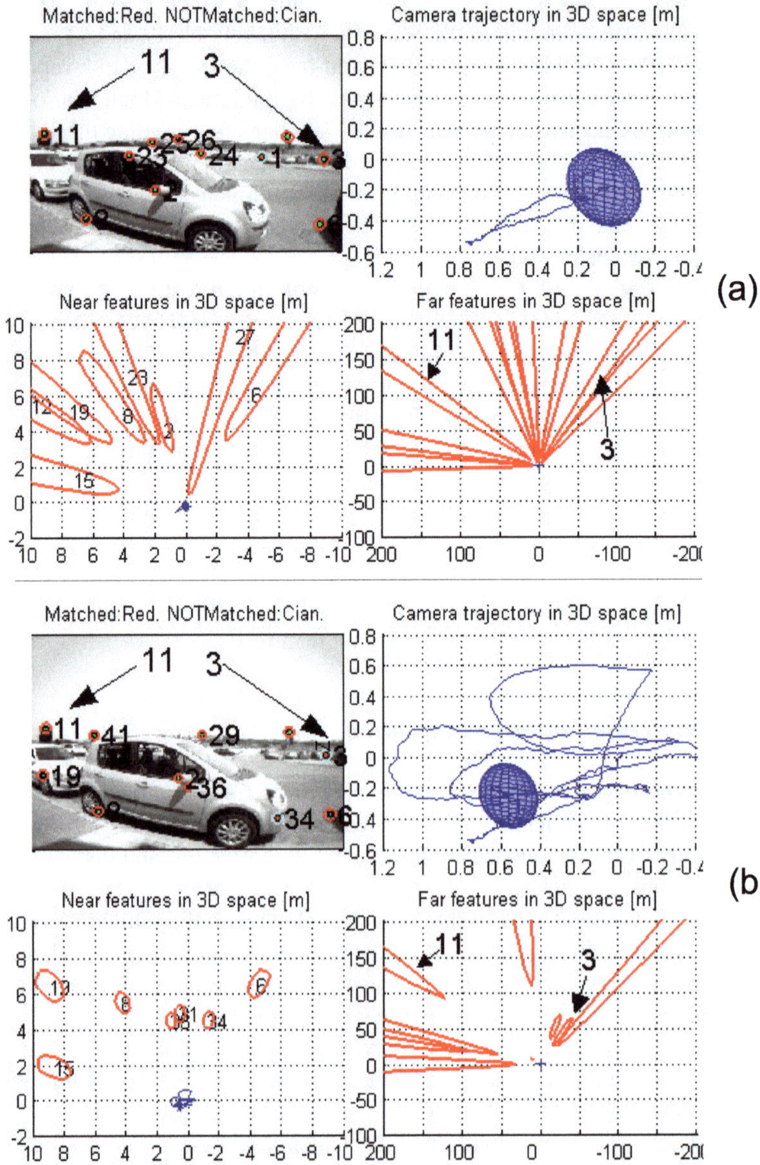

Fig. 3.8. Subfigures (a) and (b) show frames #163 and #807 from the outdoor experiment of section 3.8.2. This experiment was processed in real time. The focus was two features: 11 (tree on the left) and 3 (car on the right) at low parallax. Each of the two subfigures shows the current images, and top-down views illustrating the horizontal components of the estimation of camera and feature locations at three different zoom scales for clarity: the top-right plots (maximum zoom) highlight the estimation of the camera motion; bottom-left (medium zoom) views highlight nearby features; and bottom-right (minimum zoom) emphasizes distant features.

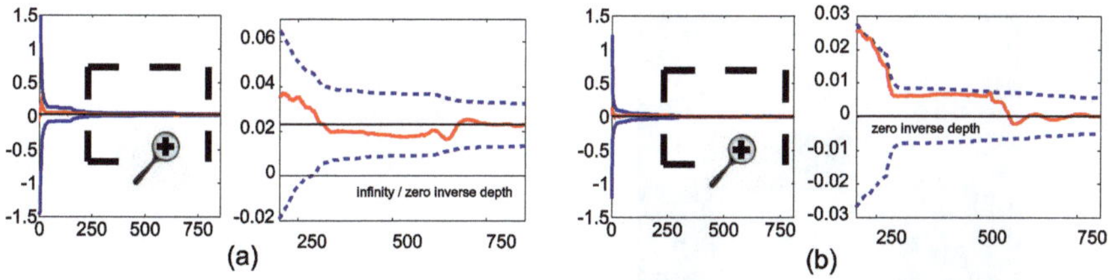

Fig. 3.9. Analysis of outdoor experiment of section 3.8.2. (a) Inverse depth estimation history for feature 3, on the car, and (b) for feature 11, on a distant tree. Due to the uncertainty reduction during estimation, two plots at different scales are shown for each feature. It is show the 95% confidence region, and with a thick line the estimated inverse depth. The thin solid line is the inverse depth estimated after processing the whole sequence. In (a), for the first 250 steps, zero inverse depth is included in confidence region, meaning that the feature might be at infinity. After this, more distant but finite locations are gradually eliminated, and eventually the feature's depth is accurately estimated. In (b), the tree is so distant that the confidence region always includes zero, since little parallax is gathered for that feature.

this feature without obtaining any information from it, while in our system it behaves like a bearing reference. This ability to deal with distant points in real time is a highly advantageous quality of our parametrization. Note that what does happen to the estimate of Feature 11 as translation occurs is that hypotheses of nearby depths are ruled out — the inverse depth scheme correctly recognizes that measuring little parallax while the camera has translated some distance allows a minimum depth for the feature to be set.

3.8.3 Loop Closing Sequence

A loop closing sequence offers a challenging benchmark for any SLAM algorithm. In this experiment a hand-held camera was carried by a person walking in small circles within a very large student laboratory, carrying out two complete laps.

Figure 3.10 shows a selection of the 737 frames from the sequence, concentrating on the beginning, first loop closure and end of the sequence. Figure 3.11 shows the camera location estimate covariance history, represented by the 95% confidence regions for the 6 camera d.o.f. and expressed in a reference local to the camera.

We observe the following properties of the evolution of the estimation, focussing in particular on the uncertainty in the camera location:

- After processing the first few images, the uncertainty in the features' depth is still huge, with highly non-elliptical confidence regions in XYZ space (Figure 3.10(a)).
- In Figure 3.11 the first peak in the X and Z translation uncertainty corresponds to a camera motion backwards along the optical axis; this motion produces poor parallax for newly initialized features, and we therefore see a reduction in

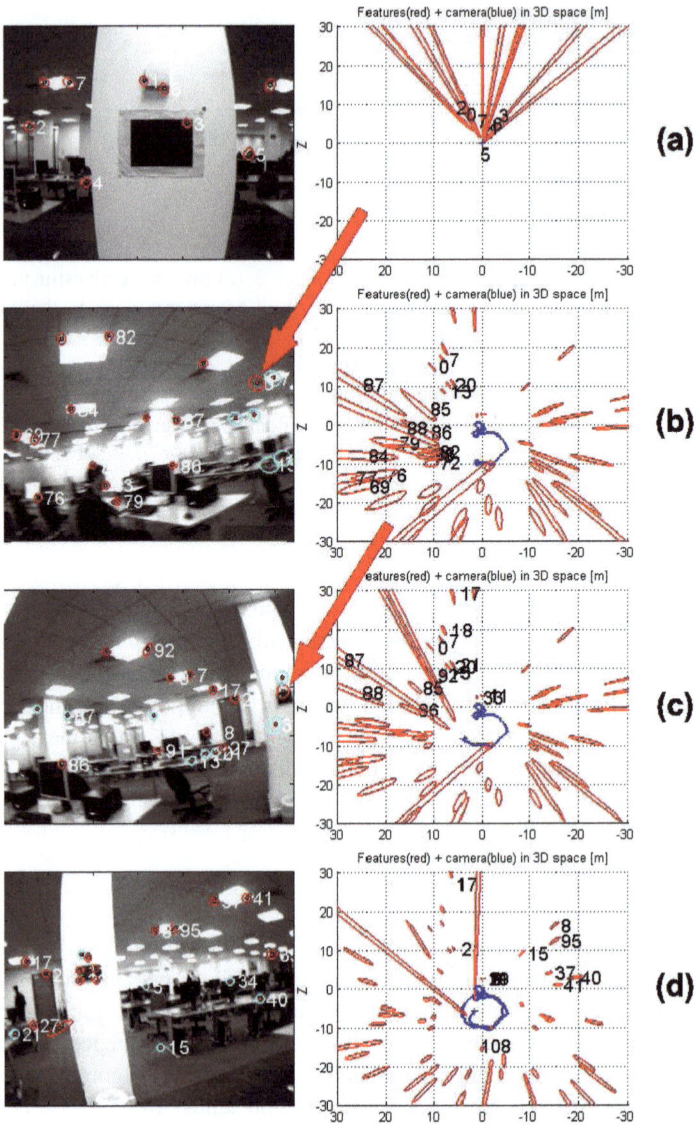

Fig. 3.10. A selection of frames from the loop closing experiment of section 3.8.3. For each frame, we show the current image and the projection of the estimated map (left), and a top-down view of the map with 95% confidence regions and camera trajectory (right). Notice that confidence regions for the map features are far from being Gaussian ellipses, especially for newly initialized or distant features. The selected frames are: (a) #11, close to the start of the sequence; (b) #417, where the first loop closing match, corresponding to a distant feature, is detected; the loop closing match is signaled with an arrow; (c) #441 where the first loop closing match corresponding to a close feature is detected; the match is signaled with an arrow; and (d) #737, the last image, in the sequence, after reobserving most of the map features during the second lap around the loop.

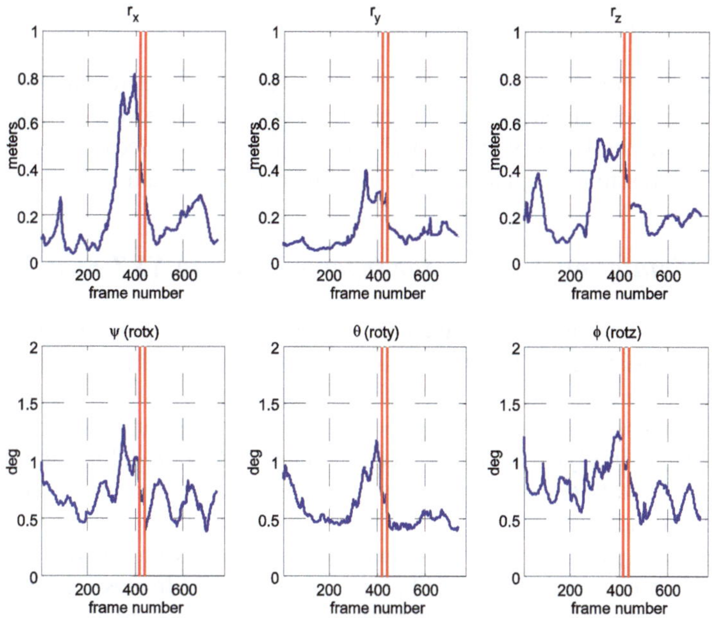

Fig. 3.11. Camera location estimate covariance along the sequence. The 95% confidence regions for each of the 6 d.o.f of camera motion are plotted. Note that errors are expressed in a reference local to the camera. The vertical solid lines indicate the loop closing frames #417 and #441.

orientation uncertainty and an increase in translation uncertainty. After frame #50 the camera again translates in the X direction, parallax is gathered and the translation uncertainty is reduced.

- From frame #240, the camera starts a 360° circular motion in the XZ plane. The camera explores new scene regions, and the covariance increases steadily as expected (Figure 3.11).
- In frame #417, the first loop closing feature is re-observed. This is a feature which is distant from the camera, and causes an abrupt reduction in orientation and translation uncertainty (Figure 3.11), though a medium level of uncertainty remains.
- In frame #441, a much closer loop closing feature (mapped with high parallax) is matched. Another abrupt covariance reduction takes place (Figure 3.11) with the extra information this provides.
- After frame #441, as the camera goes on a second lap around the loop, most of the map features are revisited, almost no new features are initalized, and hence the uncertainty in the map is further reduced. Comparing the map at frame #441 (the beginning of the second lap) and at #737, (the end of the second lap), we see a significant reduction in uncertainty. During the second lap, the camera

uncertainty is low, and as features are reobserved their uncertainties are noticeably reduced (Figure 3.10(c) and (d)).

Note that these loop closing results with the inverse depth representation show a marked improvement on the experiments on monocular SLAM with a humanoid robot presented in [Davison *et al.* 2007], where a gyro was needed in order to reduce angular uncertainty enough to close loops with very similar camera motions.

3.8.4 Simulation Analysis for Inverse Depth to XYZ Switching

In order to analyze the effect of the parametrization switching proposed in section 3.6 on the consistency of SLAM estimation, simulation experiments with different switching thresholds were run. In the simulations, a camera completed two laps of a circular trajectory of radius 3m in the XZ plane, looking out radially at a scene composed of points lying on three concentric spheres of radius 4.3m, 10m and 20m. These points at different depths were intended to produce observations with a range of parallax angles (Figure 3.12.)

The camera parameters of the simulation correspond with our real image acquisition system: camera 320×240 pixels, frame rate 30 frames/sec, image field of view 90°, measurement uncertainty for a point feature in the image, Gaussian

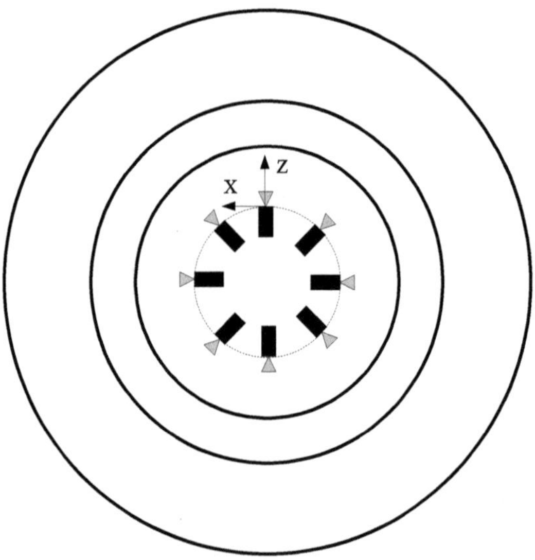

Fig. 3.12. Simulation configuration for analysis of parametrization switching in section 3.8.4, sketching the circular camera trajectory and 3D scene, composed of three concentric spheres of radius 4.3m, 10m and 20m. The camera completes two circular laps in the (XZ) plane with radius 3m, and is orientated radially.

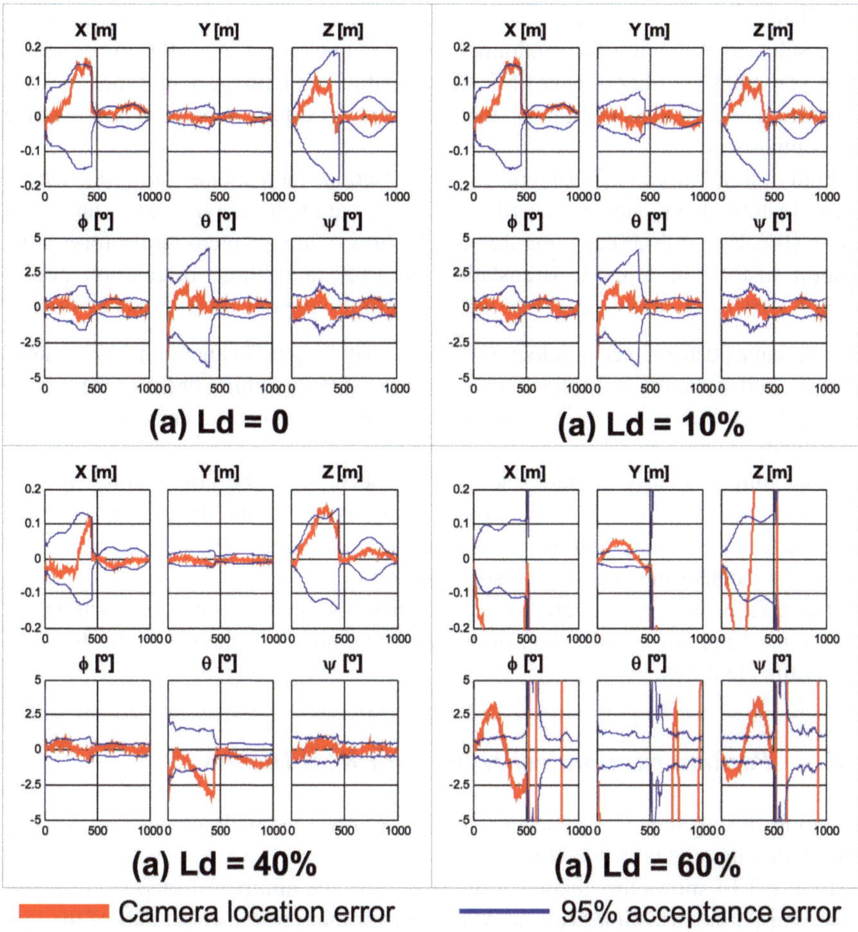

Fig. 3.13. Details from the parametrization switching experiment. Camera location estimation error history in 6 d.o.f. (translation in XYZ, and three orientation angles $\psi\theta\phi$) for four switching thresholds: With $L_d = 0\%$, no switching occurs and the features all remain in the inverse depth parametrization. At, $L_d = 10\%$ although features from the spheres at 4.3m and 10m are eventually converted, no degradation with respect to the non-switching case is observed. At $L_d = 40\%$ some features are switched before achieving true Gaussianity, and there is noticeable degradation, especially in θ rotation around the Y axis. At $L_d = 60\%$ the map becomes totally inconsistent and loop closing fails.

$N\left(0, 1\text{pixel}^2\right)$. The simulated image sequence contained 1000 frames. Features were selected following the randomized map management algorithm proposed in [Davison 2003] in order to have 15 features visible in the image at all times. All our simulation experiments work using the same scene features, in order to homogenize the comparison.

Four simulation experiments for different thresholds for switching each feature from inverse depth to XYZ parametrization were run, with $L_d \in \{0\%, 10\%, 40\%, 60\%\}$. Figure 3.13 shows the camera trajectory estimation history in 6 d.o.f. (translation in XYZ, and three orientation angles $\psi(\text{Rot}_x), \theta(\text{Rot}_y), \phi(\text{Rot}_z, \text{cyclotorsion})$). The following conclusions can be made:

- The same performance is achieved with no switching (0%) and with 10% switching. So it is clearly advantageous to perform 10% switching because there is no penalization in accuracy and the state vector size of each converted feature is halved.
- Switching too early degrades accuracy, especially in the orientation estimate. Notice how for 40% the orientation estimate is worse and the orientation error covariance is smaller, showing filter inconsistency. For 60%, the estimation is totally inconsistent and loop closing fails.
- Since early switching degrades performance, the inverse depth parametrization is mandatory for initialization of every feature and over the long-term for low-parallax features.

3.8.5 Parametrization Switching with Real Images

The loop closing sequence of section 3.8.3 was processed without any parametrization switching, and with switching at $L_d = 10\%$.

Figure 3.14 shows the history of the state size, the number of map features and how their parametrization evolves. At the last estimation step about half of the features had been switched; at this step the state size had reduced from 427 to 322 (34 inverse depth features and 35 XYZ), i.e. 75% of the original vector size. Figure 3.15 shows four frames from the sequence illustrating feature switching. Up to step 100 the camera has low translation and all the features are in inverse depth form. As the camera translates nearby features switch to XYZ. Around step 420, the loop is closed and features are reobserved, producing a significant reduction in uncertainty which allows switching of more reobserved close features. Our method automatically determines which features should be represented in the inverse depth or XYZ forms, optimizing computational efficiency without sacrificing accuracy.

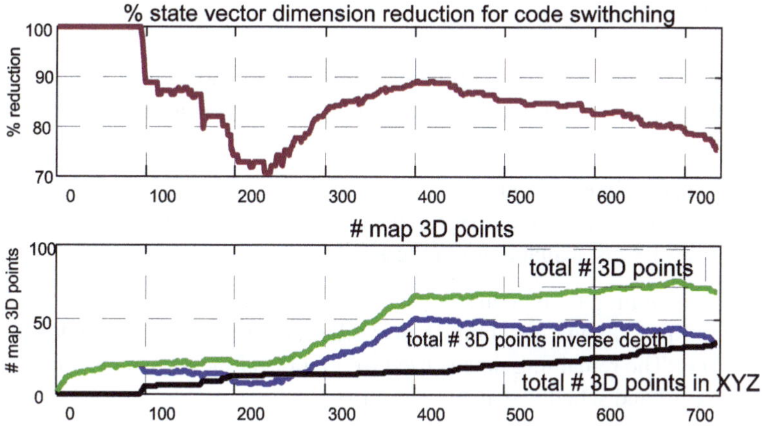

Fig. 3.14. Parametrization switching on a real sequence (section 3.8.5): state vector size history. Top: percentage reduction in state dimension when using switching compared with keeping all points in inverse depth. Bottom: total number of points in the map, showing the number of points in inverse depth and the number of points in XYZ.

Fig. 3.15. Parametrization switching seen in image space: points coded in inverse depth ($\star$) and coded in XYZ ($\triangle$). (a) First frame, with all features in inverse depth. (b) Frame #100; nearby features start switching. (c) Frame # 470, loop closing; most features in XYZ. (d) Last image of the sequence.

3.8.6 *Processing Time*

We give some details of the real-time operation of our monocular SLAM system, running on a 1.8 GHz. Pentium M processor laptop. A typical EKF iteration would imply:

- A state vector dimension of 300.
- 12 features observed in the image, a measurement dimension of 24.
- 30 fps, so 33.3 ms available for processing.

Typical computing time breaks down as follows: Image acquisition, 1 ms.; EKF prediction, 2 ms.; Image matching, 1 ms.; EKF update, 17 ms. That adds up to a total of 21ms. The remaining time is used for graphics functions, using OpenGL on an NVidia card and scheduled at a low priority.

The quoted state vector size 300 corresponds to a map size of 50 if all features are encoded using inverse depth. In indoor scenes, thanks to switching maps of up to 60-70 features can be computed in real time. This size is enough to map many typical scenes robustly.

3.9 Discussion

We have presented a parametrization for monocular SLAM which permits operation based uniquely on the standard EKF prediction-update procedure at every step, unifying initialization with the tracking of mapped features. Our inverse depth parametrization for 3D points allows unified modelling and processing for any point in the scene, close or distant, or even at 'infinity'. In fact, close, distant or just-initialized features are processed within the routine EKF prediction-update loop without making any binary decisions. Thanks to the undelayed initialization and immediate full use of infinite points, estimates of camera orientation are significantly improved, reducing the camera estimation jitter often reported in previous work. The jitter reduction in turn leads to computational benefits in terms of smaller search regions and improved image processing speed

The key factor is that due to our parametrization of the direction and inverse depth of a point relative to the location from which it was first seen, our measurement equation has low linearization errors at low parallax, and hence the estimation uncertainty is accurately modeled with a multi-variate Gaussian. In section 3.4 we presented a model which quantifies linearization error. This provides a theoretical understanding of the impressive outdoor, real-time performance of the EKF with our parametrization.

The inverse depth representation requires a six-dimensional state vector per feature, compared to three for XYZ coding. This doubles the map state vector size, and hence produces a 4-fold increase in the computational cost of the EKF. Our experiments show that it is essential to retain the inverse depth parametrization for intialization and distant features, but that nearby features can be safely converted to the cheaper XYZ representation meaning that the long-term computational cost need not significantly increase. We have given details on when this conversion should be carried out for each feature, to optimize computational efficiency without sacrificing accuracy.

The experiments presented have validated the method with real imagery, using a hand-held camera as the only sensor both indoors and outdoors. We have experimentally verified the key contributions of our work:

- Real-time performance achieving 30 fps real-time processing for maps up to 60–70 features.
- Real-time loop closing.
- Dealing simultaneously with low and high parallax features.
- Non delayed initialization.
- Low jitter, full 6 DOF monocular SLAM.

The point parametrization detailed in this chapter –first presented in [Montiel *et al.* 2006, Civera *et al.* 2007a, Civera *et al.* 2008a]– has reached the status of *standard* and reached an impact in the robotics community that is worth summarizing here. The inverse depth parametrization is used in the most successful implementations of EKF-based monocular SLAM [Clemente *et al.* 2007, Williams *et al.* 2007, Piniés & Tardós 2008, Paz *et al.* 2008b, Holmes *et al.* 2008, Castle *et al.* 2010]. Several modifications to the approach presented here has been studied [Marzorati *et al.* 2008, Solà 2010]. Nevertheless, it has been recently proved in [Solà *et al.* 2011] that the one presented here still presents the highest degree of accuracy and consistency. The inverse depth idea has been also succesfully applied to line-based monocular SLAM [Solà *et al.* 2009], stereo SLAM [Paz *et al.* 2008a] and SLAM with an omnidirectional camera [Rituerto *et al.* 2010].

Chapter 4
1-Point RANSAC

Abstract. Random Sample Consensus (RANSAC) has become one of the most successful techniques for robust estimation from a data set that may contain outliers. It works by constructing model hypotheses from random minimal data subsets and evaluating their validity from the support of the whole data. In this chapter we present an efficient RANSAC algorithm for an Extended Kalman Filter (EKF) framework that uses the available prior probabilistic information from the EKF in the RANSAC model hypothesize stage. This allows the minimal sample size to be reduced to one, resulting in large computational savings without performance degradation. 1-Point RANSAC is also shown to outperform both in accuracy and computational cost the Joint Compatibility Branch and Bound (JCBB) algorithm, a gold-standard technique for spurious rejection within the EKF framework. The combination of this 1-point RANSAC and the robocentric formulation of the EKF SLAM allows a qualitative jump on the general performance of the algorithms presented in this book: In this chapter, sequences covering trajectories of several hundreds of metres are processed showing highly accurate camera motion estimation results.

4.1 Introduction

The establishment of reliable correspondences from sensor data is at the core of most estimation algorithms in robotics. The search for correspondences, or data association, is usually based first stage on comparing local descriptors of salient features in the measured data. The ambiguity of such local description usually produces incorrect correspondences at this stage. Robust methods operate by checking the consistency of the data against the global model assumed to be generating the data, and discarding as spurious any that does not fit into it. Among robust estimation methods, Random Sample Consensus (RANSAC) [Fischler & Bolles 1981] stands out as one of the most successful and widely used, especially in the Computer Vision community.

J. Civera et al.: Structure from Motion Using the Extended Kalman Filter, STAR 75, pp. 65–97.
springerlink.com

This chapter introduces a novel integration of RANSAC into the Extended Kalman Filter framework. In order to highlight the requirements and benefits of this method, the RANSAC algorithm is first briefly exposed in this introduction for the simple case of 2D line estimation from a set of points contaminated with spurious data (see Figure 4.1). After that, the same simple example will be tackled using the proposed 1-Point RANSAC algorithm (Figure 4.2). It is important to remark here that we use this simple example only to illustrate in the simplest manner our approach, and will later on fill in the details which make 1-Point RANSAC into a fully practical matching algorithm.

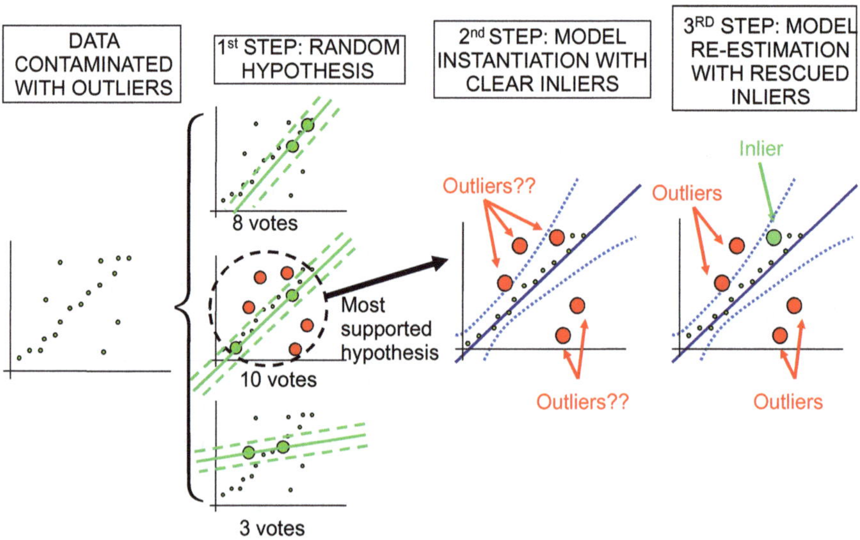

Fig. 4.1. RANSAC steps for the simple 2D line estimation example: First, random hypotheses are generated from data samples of size two, the minimum to define a line. The most supported one is selected, and data voting for this hypothesis is considered inlier. Model parameters are estimated from those clear inliers in a second step. Finally, the remaining data points consistent with this latest model are rescued and the model is re-estimated again.

Standard RANSAC starts from a set of data, in our simple example 2D points, and the underlying model that generates the data, a 2D line. In the first step, RANSAC constructs hypotheses for the model parameters and selects the one that gathers most support. Hypotheses are randomly generated from the minimum number of points necessary to compute the model parameters, which is two in our case of line estimation. Support for each hypothesis can be computed in its most simple form by counting the data points inside a threshold (related to the data noise), although more sophisticated methods have been used [Torr & Zisserman 2000].

Hypotheses involving one or more outliers are assumed to receive low support, as is the case in the third hypothesis in Figure 4.1. The number of hypotheses n_{hyp} necessary to ensure that at least one spurious-free hypothesis has been tested with probability p can be computed from this formula:

$$n_{hyp} = \frac{log(1-p)}{log(1-(1-\varepsilon)^m)}, \quad (4.1)$$

where ε is the outlier ratio and m the minimum number of data points necessary to instantiate the model. The usual approach is to adaptively compute this number of hypotheses at each iteration, assuming the inlier ratio is the support set by the total number of data points in this iteration [Hartley & Zisserman 2004].

Data points that voted for the most supported hypothesis are considered clear inliers. In a second stage, clear inliers are used to estimate the model parameters. Individual compatibility is checked for each one of the rest of the points against the estimated model. If any of them is rescued as inlier, as happens in the example in Figure 4.1, the model parameters are re-estimated again in a third step.

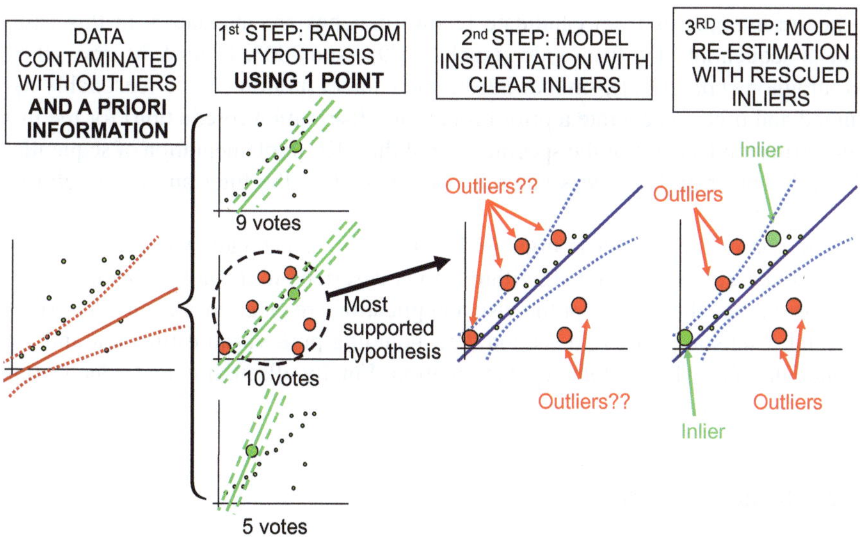

Fig. 4.2. 1-Point RANSAC steps for the simple 2D line estimation example: As a key difference from standard RANSAC, the algorithm assumes that an a priori probability distribution over the model parameters is known in advance. This prior knowledge allows us to compute the random hypotheses using only 1 data point, hence reducing the number of hypotheses and the computational cost. The remaining steps do not vary with respect to standard RANSAC in Figure 4.1

Figure 4.2 illustrates the idea behind 1-Point RANSAC in the same 2D line estimation problem. As the first key difference, the starting point is a data set and its underlying model, but also a prior probability distribution over the model parameters. RANSAC random hypotheses are then generated based on this prior information and data points, differently from standard RANSAC hypothesis solely based on data points. The use of prior information can reduce the size of the data set that instantiates the model to the minimum size of one point, and it is here where the computational benefit of our method with respect to RANSAC arises: according to Equation 4.1, reducing the sample size m greatly reduces the number of RANSAC iterations and hence the computational cost.

The order of magnitude of this reduction can be better understood if we switch from the simple 2D line estimation example to our visual estimation application. According to [Nistér 2004], at least five image points are necessary to estimate the 6 degrees of freedom camera motion between two frames (so $m = 5$). Using formula 4.1, assuming an inlier ratio of 0.5 and a probability p of 0.99, the number of random hypotheses would be 146. Using our 1-Point RANSAC scheme, assuming that probabilistic a priori information is available, the sample size m can be reduced to one point and the number of hypotheses would be reduced to 7.

Having an a priori probability distribution over the camera parameters is unusual in classical pairwise Structure from Motion which assumes widely separated views [Hartley & Zisserman 2004], and methods like standard RANSAC which generate hypotheses from candidate feature matches are mandatory in this case. But in sequential SfM from video (such as [Davison 2003, Klein & Murray 2008, Mouragnon *et al.* 2009]), smooth interframe camera motion can be reasonably assumed and used to generate a prior probability distribution (prediction) for the image correspondences. For the specific case of the EKF implementation of sequential SfM, this prior probability is naturally propagated by the filter and is straightforwardly available.

The rest of the chapter is organised as follows: first, related work is described in section 4.2; the proposed algorithm is then described in its most general form in section 4.3 and the details for the visual application are given in section 4.4. Experimental results are shown in section 4.5, including pure visual estimation and the monocular and wheel odometry combination. Finally, discussion and conclusions are presented in sections 4.6 and **??**.

4.2 Related Work

4.2.1 Random Sample Consensus (RANSAC)

Although RANSAC is a relatively old method, the literature covering the topic continues up to the present. RANSAC [Fischler & Bolles 1981] was introduced early in visual geometric estimation [Torr & Murray 1993] and has been the preferred outlier rejection tool in the field. Recently, an important stream of research has focused on reducing the model verification cost in standard RANSAC (e.g. [Raguram *et al.* 2008, Chum & Matas 2008, Capel 2005, Nistér 2005]) via the early detection and termination of bad hypotheses. The 1-point RANSAC algorithm proposed here is related to this stream in the sense that it also reduces the hypothesis generation and validation cost. Nevertheless, it does so in a different manner: instead of fast identification of good hypotheses among a large number of them, the number of hypotheses is greatly reduced from the start by considering the prior information given by a dynamic model.

Incorporating probabilistic information into RANSAC has rarely been discussed in the computer vision literature. Only very recently Moreno *et al.* [Moreno-Noguer *et al.* 2008] have explored the case where weak a priori information is available in the form of probabilistic distribution functions.

More related to our method, the combination of RANSAC and Kalman filtering was proposed by Vedaldi *et al.* in [Vedaldi *et al.* 2005]. Our 1-Point RANSAC might be considered as a specific form of Vedaldi's quite general approach. They propose an iterative scheme in which several minimal hypotheses are tested; for each such hypothesis all the consistent matches are iteratively harvested. No statement about the cardinality of the hypotheses is made. Here we propose a definite and efficient method, in which the cardinality of the hypotheses generator size is 1, and the inlier harvesting is not iterative but in two stages. Finally we describe in reproducible detail how to deal efficiently with the EKF algorithm in order to reach real-time, splitting the expensive EKF covariance update in two stages.

RANSAC using 1-point hypotheses has also been very recently proposed in [Scaramuzza *et al.* 2009] as the result of constraining the camera motion. While at least 5 points would be needed to compute monocular Structure from Motion for a calibrated camera undergoing general six degrees of freedom motion [Nistér 2004], fewer are needed if the motion is known to be less general: as few as 2 points in [Ortín & Montiel 2001] for planar motion and 1 point in [Scaramuzza *et al.* 2009] for planar and nonholonomic motion. As a clear limitation of both approaches, any motion performed out of the model will result in estimation error. In fact, it is shown in real-image experiments in [Scaramuzza *et al.* 2009] that although the most constrained model is enough for RANSAC hypotheses (reaching then 1-point RANSAC), a less restrictive model offers better results for motion estimation.

In the case of the new 1-point RANSAC presented here, extra information for the predicted camera motion comes from the probability distribution function that the EKF naturally propagates over time. The method presented is then in principle not restricted to any specific motion, being suitable for 6 degrees of freedom estimation. The only assumption is the existence of tight and highly correlated priors, which is reasonable within the EKF framework since the filter itself only works in such circumstances.

4.2.2 *Joint Compatibility Branch and Bound (JCBB)*

Joint Compatibility Branch and Bound (JCBB) [Neira & Tardós 2001] has been the preferred technique for spurious match rejection within the EKF framework in the robotics community, being successfully used in visual (e.g. [Clemente *et al.* 2007], [Williams *et al.* 2007]) and non-visual SLAM (e.g. [Fenwick *et al.* 2002]). Unlike RANSAC, which hypothesizes model parameters based on current measurement data, JCBB detects spurious measurements based on a predicted probability distribution over the measurements. It does so by extracting from all the possible matches the maximum set that is jointly compatible with the multivariate Gaussian prediction.

In spite of its wide use, JCBB presents two main limitations that 1-Point RANSAC overcomes. First, JCBB operates over the prediction for the measurements *before* fusing them. Such a probabilistic prediction is coming from the linearization of the dynamic and measurement models and the assumption of Gaussian noise; so it will presumably not correspond to the real state of the system. 1-Point and in general any RANSAC operates over hypotheses *after* the integration of a data subset, which have corrected part of the predicted model error with respect to the real system.

The second limitation of JCBB concerns computational cost: the Branch and Bound search that JCBB uses for extracting the largest jointly compatible set of matches has exponential complexity in the number of matches. This complexity does not present a problem for small numbers of matches, as is the case in the references two paragraphs above, but very large computation times arise when the number of spurious grows, as we will show in the experimental results section. The computational complexity of 1-Point RANSAC is linear in the state and measurement size and exhibits low cost variation with the number of outliers.

Two recent methods are also of interest for this work. First, Active Matching [Chli & Davison 2008] is a clear inspiration for our method. In Active Matching, feature measurements are integrated sequentially, with the choice of measurement at each step driven by expected information gain, and the results of each measurement in turn used to narrow the search for subsequent correspondences. 1-Point RANSAC can be seen as lying in the middle ground between RANSAC and JCBB which obtain point correspondence candidates and then aim to resolve them, and Active Matching with its fully sequential search for correspondence. The first step of 1-Point RANSAC is very similar to Active Matching, and confirming that integrating the first match highly constrains the possible image locations of other features, but afterwards the methods of the algorithms diverge. A problem with Active Matching in [Chli & Davison 2008] was the unreasonably high computational cost of scaling to large numbers of feature correspondences per frame, and 1-Point RANSAC has much better properties in this regard, though very recently an improvement to Active Matching has also addressed this issue in a different way [Handa *et al.* 2010].

Paz *et al.* [Paz *et al.* 2008b] describe an approach called Randomized Joint Compatibility (RJC) which basically randomizes the jointly compatible set search, avoiding the Branch and Bound search and ensuring an initial small set of jointly compatible inliers at the first step via Branch and Bound search in random sets. Only afterwards, the joint compatibility of each remaining match is checked against the initial set. Although this approach lowers the computational cost of the JCBB, it still faces the accuracy problems derived from the use of the predicted measurement function before data fusion.

4.2.3 *Structure from Motion and Visual Odometry*

Structure from Motion (SfM) is the generic term for 3D estimation from the sole input of a set of images of the imaged 3D scene and the corresponding camera locations. SfM from a sparse set of images has been usually processed by pairwise geometry algorithms [Hartley & Zisserman 2004] and refined by global optimization procedures [Triggs *et al.* 2000]. Estimation from a sequence has been carried out either by local optimization of keyframes [Klein & Murray 2008, Mouragnon *et al.* 2009], or by filtering [Davison *et al.* 2007, Eade & Drummond 2007].

Visual Odometry, a term coined in [Nistér *et al.* 2004], refers to egomotion estimation mainly from visual input (monocular or stereo), but sometimes also combined with mechanical odometry and/or inertial sensor measurements. The variety of approaches here makes a complete review difficult; some visual odometry algorithms have made use of stereo cameras, either as the only sensor (e.g. [Comport *et al.* 2007]) or in combination with inertial measurements [Konolige *et al.* 2007, Cheng *et al.* 2006]. Among the monocular approaches, [Mouragnon *et al.* 2009] uses a non-panoramic camera as the only sensor. Several others have been proposed using an omnidirectional camera, e.g. [Scaramuzza *et al.* 2009, Tardif *et al.* 2008]. The experiment presented here, combining a non-panoramic camera plus proprioceptive information for estimation of large trajectories, is rarely found in the literature.

4.2.4 *Benchmarking*

Carefully designed benchmark datasets and methods have come into standard use in the vision community, e.g. [Scharstein & Szeliski 2002, Baker *et al.* 2007, Everingham *et al.* 2010]. Robotics datasets have only recently reached such level of detail, presenting either detailed benchmarking procedures [Kummerle *et al.* 2009] or datasets with reliable ground truth and open resources for comparison [Smith *et al.* 2009, Blanco *et al.* 2009].

The RAWSEEDS dataset [RAWSEEDS 2011], which include monocular and wheel odometry streams for large scale scenarios, will be used for the Visual Odometry experiments in the next section of this chapter. While being suitable to benchmark very large real-image experiments, robotic datasets face two main inconveniences: First, the robot motion is planar in all the datasets, thus not allowing to evaluate full six-degrees-of-freedom motion estimation. And second, GPS only provides translational data and angular estimation cannot be benchmarked. Simulation environments, like the one described in [Funke & Pietzsch 2009], can provide the translational and angular ground truth for any kind of camera motion. Nevertheless, those simulation environments usually cannot represent full real world complexity.

The benchmarking method proposed and used in this chapter overcomes all these limitations. It consists of comparing the estimation results against a Bundle Adjustment solution over high resolution images. Full 6 DOF motion can be evaluated with low user effort (only the generation of a Bundle Adjustment solution is required), requirements for hardware are low (a high resolution camera) and any kind of motion or scene can be evaluated as the method operates over the real images themselves.

This approach is not entirely new: the use of a global Bundle Adjustment solution to benchmark sequential algorithms has already been used in [Eade & Drummond 2007, Mouragnon *et al.* 2009]. The contribution here is the validation of the algorithm, effectively showing that the Bundle Adjustment uncertainty is much lower than the sequential methods to benchmark. As another novelty, global Bundle Adjustment is applied over high resolution images, further improving accuracy. While it is true that a Bundle Adjustment solution still may suffer from scale drift, it will be much lower than that of the sequential algorithms. Also, scale drift can be driven close to zero by carefully chosing the images over which to apply Bundle Adjustment to form a well-conditioned network [Triggs *et al.* 2000], so the validity of the method is not compromised.

4.3 1-Point RANSAC Extended Kalman Filter Algorithm

Algorithm 1 outlines the proposed combination of 1-Point RANSAC inside the EKF framework in its most general form, and we describe this in detail in this section. The language used here is deliberately general in the belief that the described algorithm may be of application in a large number of estimation problems. The particular scenarios of the experimental results section (real-time sequential visual odometry from a monocular sequence, either with or without additional wheel odometry) are discussed in detail in section 4.4.

4.3.1 EKF Prediction and Search for Individually Compatible Matches (lines 5–8)

The algorithm begins with standard EKF prediction: the estimation for the state vector $\mathbf{x}_{k-1|k-1}$ at step $k-1$, modeled as a multidimensional Gaussian $\mathbf{x}_{k-1|k-1} \sim \mathcal{N}\left(\hat{\mathbf{x}}_{k-1|k-1}, \mathbf{P}_{k-1|k-1}\right)$, is propagated to step k through the known dynamic model $\mathbf{f}_k$

$$\hat{\mathbf{x}}_{k|k-1} = \mathbf{f}_k\left(\hat{\mathbf{x}}_{k-1|k-1}, \mathbf{u}_k\right) \tag{4.2}$$

$$\mathbf{P}_{k|k-1} = \mathbf{F}_k \mathbf{P}_{k-1|k-1} \mathbf{F}_k^\top + \mathbf{G}_k \mathbf{Q}_k \mathbf{G}_k^\top \, . \tag{4.3}$$

In the above equation $\mathbf{u}_k$ stands for the control inputs to the system at step k, $\mathbf{F}_k$ is the Jacobian of $\mathbf{f}_k$ with respect to the state vector $\mathbf{x}_{k|k-1}$ at step k, $\mathbf{Q}_k$ is the covariance of the zero-mean Gaussian noise assumed for the dynamic model and $\mathbf{G}_k$ is the Jacobian of the state vector $\mathbf{x}_{k|k-1}$ by the input $\mathbf{u}_k$ at step k.

Algorithm 1. 1-Point RANSAC EKF

INPUT: $\hat{\mathbf{x}}_{k-1|k-1}, \mathbf{P}_{k-1|k-1}$ {EKF estimate at step $k-1$}
th {Threshold for low-innovation points, related with the measurement noise.}
OUTPUT: $\hat{\mathbf{x}}_{k|k}, \mathbf{P}_{k|k}$ {EKF estimate at step k}

{A. EKF prediction and individually compatible matches}
$[\hat{\mathbf{x}}_{k|k-1}, \mathbf{P}_{k|k-1}] = EKF\_prediction(\hat{\mathbf{x}}_{k-1|k-1}, \mathbf{P}_{k-1|k-1}, \mathbf{u})$
$[\hat{\mathbf{h}}_{k|k-1}, \mathbf{S}_{k|k-1}] = measurement\_prediction(\hat{\mathbf{x}}_{k|k-1}, \mathbf{P}_{k|k-1})$
$\mathbf{z}^{IC} = search\_IC\_matches(\hat{\mathbf{h}}_{k|k-1}, \mathbf{S}_{k|k-1})$

{B. 1-Point hypotheses generation and evaluation}
$\mathbf{z}^{li\_inliers} = [\,]$
$n_{hyp} = 1000$ {Initial value, will be updated in the loop}
for $i = 0$ to n_{hyp} **do**
  $\mathbf{z}_i = select\_random\_match(\mathbf{z}^{IC})$
  $\hat{\mathbf{x}}_i = EKF\_state\_update(\mathbf{z}_i, \hat{\mathbf{x}}_{k|k-1})$ {Notice: only state update; NO covariance update}
  $\hat{\mathbf{h}}_i = predict\_all\_measurements(\hat{\mathbf{x}}_i)$
  $\mathbf{z}_i^{th} = find\_matches\_below\_a\_threshold(\mathbf{z}^{IC}, \hat{\mathbf{h}}_i, th)$
  if $size(\mathbf{z}_i^{th}) > size(\mathbf{z}^{li\_inliers})$ **then**
    $\mathbf{z}^{li\_inliers} = \mathbf{z}_i^{th}$
    $\varepsilon = 1 - \frac{size(\mathbf{z}^{li\_inliers})}{size(\mathbf{z}^{IC})}$
    $n_{hyp} = \frac{log(1-p)}{log(1-(1-\varepsilon))}$ {From equation 4.1 with $m = 1$}
  end if
end for

{C. Partial EKF update using low-innovation inliers}
$[\hat{\mathbf{x}}_{k|k}, \mathbf{P}_{k|k}] = EKF\_update(\mathbf{z}^{li\_inliers}, \hat{\mathbf{x}}_{k|k-1}, \mathbf{P}_{k|k-1})$

{D. Partial EKF update using high-innovation inliers}
$\mathbf{z}^{hi\_inliers} = [\,]$
for every match $\mathbf{z}^j$ above a threshold th **do**
  $[\hat{\mathbf{h}}^j, \mathbf{S}^j] = point\_j\_prediction\_and\_covariance(\hat{\mathbf{x}}_{k|k}, \mathbf{P}_{k|k}, j)$
  $\nu^j = \mathbf{z}^j - \hat{\mathbf{h}}^j$
  if ${\nu^j}^\top {\mathbf{S}^j}^{-1} \nu^j < \chi^2_{2,0.01}$ **then**
    $\mathbf{z}^{hi\_inliers} = add\_match\_j\_to\_inliers(\mathbf{z}^{hi\_inliers}, \mathbf{z}^j)$ {If individually compatible, add to inliers}
  end if
end for
if $size(\mathbf{z}^{hi\_inliers}) > 0$ **then**
  $[\hat{\mathbf{x}}_{k|k}, \mathbf{P}_{k|k}] = EKF\_update(\mathbf{z}^{hi\_inliers}, \hat{\mathbf{x}}_{k|k}, \mathbf{P}_{k|k})$
end if

The predicted probability distribution for the state $\mathbf{x}_{k|k-1}$ can be used to ease the correspondence search as described in section 3.7.2. Propagating this predicted state through the measurement model $\mathbf{h}_i$ offers a Gaussian prediction for each measurement:

$$\hat{\mathbf{h}}_i = \mathbf{h}_i\left(\hat{\mathbf{x}}_{k|k-1}\right) \tag{4.4}$$

$$\mathbf{S}_i = \mathbf{H}_i \mathbf{P}_{k|k-1} \mathbf{H}_i^\top + \mathbf{R}_i \,, \tag{4.5}$$

where $\mathbf{H}_i$ is the Jacobian of the measurement $\mathbf{h}_i$ with respect to the state vector $\mathbf{x}_{k|k-1}$ and $\mathbf{R}_i$ is the covariance of the Gaussian noise assumed for each individual measurement. The actual measurement $\mathbf{z}_i$ should be exhaustively searched for inside the 99% probability region defined by its predicted Gaussian $\mathcal{N}\left(\hat{\mathbf{h}}_i, \mathbf{S}_i\right)$ by comparison of the chosen local feature descriptor.

Active search allows computational savings and also constraints the matches to be individually compatible with the predicted state $\mathbf{x}_{k|k-1}$. Nevertheless, ensuring geometric compatibility for each separated match $\mathbf{z}_i$ does not guarantee the global consensus of the whole set. So, still the joint compatibility of the data against a global model has to be checked for the set individually compatible matches $\mathbf{z}^{IC} = \left(\mathbf{z}_1 \ldots \mathbf{z}_i \ldots \mathbf{z}_n\right)^\top$ previous to the EKF update.

4.3.2 1-Point Hypotheses Generation and Evaluation (lines 9–22)

Following the principles of RANSAC, random state hypotheses $\hat{\mathbf{x}}_i$ are generated and data support is computed by counting measurements inside a threshold related to the measurement noise. It is assumed here that the predicted measurements are highly correlated, such that every hypothesis computed from one match reduces most of the common uncertainty producing an inlier uncertainty close to the measurement noise.

As the key difference with respect to standard RANSAC, random hypotheses will be generated not only based on the data $\mathbf{z}^{IC} = \left(\mathbf{z}_1 \ldots \mathbf{z}_i \ldots \mathbf{z}_n\right)^\top$ but also on the predicted state $\mathbf{x}_{k|k-1} \sim \mathcal{N}\left(\hat{\mathbf{x}}_{k|k-1}, \mathbf{P}_{k|k-1}\right)$. Exploiting this prior knowledge allows us to reduce the sample size necessary to instantiate the model parameters from the minimal size to define the degrees of freedom of the model to only one data point. The termination criteria of the RANSAC algorithm in equation 4.1 grows exponentially with the sample size, so we can achieve a great reduction in the number of hypotheses.

Another key aspect for the efficiency of the algorithm is that each hypothesis $\hat{\mathbf{x}}_i$ generation only needs an EKF state update using a single match $\mathbf{z}_i$. A covariance update, which is of quadratic complexity in the size of the state, is not needed and the cost per hypothesis will be low. Hypothesis support is calculated by projecting the updated state into the camera, which can also be performed at very low cost compared with other stages in the EKF algorithm.

4.3.3 Partial Update with Low-Innovation Inliers (lines 23–24)

Data points voting for the most supported hypothesis $\mathbf{z}^{li\_inliers}$ are designated as low-innovation inliers. They are assumed to be generated by the true model, as they are at a small distance from the most supported hypothesis. The rest of the points can be outliers but also inliers, even if they are far from the most supported hypothesis.

A simple example related to this book can illustrate this: it is well known that distant points are useful for estimating camera rotation while close points are necessary to estimate translation; as discussed in chapter 3. In the RANSAC hypotheses generation step, a distant feature would generate a highly accurate 1-point hypothesis for rotation, while translation would remain inaccurately estimated. Other distant points would in this case have low innovation and would vote for this hypothesis. But as translation is still inaccurately estimated, nearby points would presumably exhibit high innovation even if they are inliers.

So after having determined the most supported hypothesis and the other points that vote for it, some inliers still have to be "rescued" from the high-innovation set. Such inliers will be rescued after a partial state and covariance update using only the reliable set of low-innovation inliers:

$$\hat{\mathbf{x}}_{k|k} = \hat{\mathbf{x}}_{k|k-1} + \mathbf{K}' \left(\mathbf{z}^{li\_inliers} - \mathbf{h}' \left(\hat{\mathbf{x}}_{k|k-1} \right) \right) \tag{4.6}$$

$$\mathbf{P}_{k|k} = \left(\mathbf{I} - \mathbf{K}'\mathbf{H}' \right) \mathbf{P}_{k|k-1} \tag{4.7}$$

$$\mathbf{K}' = \mathbf{P}_{k|k-1} \mathbf{H}'^{\top} \left(\mathbf{H}' \mathbf{P}_{k|k-1} \mathbf{H}'^{\top} + \mathbf{R}' \right)^{-1} . \tag{4.8}$$

$\mathbf{H}' = \left(\mathbf{H}'_1 \ldots \mathbf{H}'_i \ldots \mathbf{H}'_n \right)^{\top}$ stands for the Jacobian of the measurement equation $\mathbf{h}' \left(\hat{\mathbf{x}}_{k|k-1} \right)$ that projects the low-innovation inliers into the sensor space. $\mathbf{R}'$ is the covariance assigned to the sensor noise.

4.3.4 Partial Update with High-Innovation Inliers (lines 25–35)

After a partial update using low-innovation inliers, most of the correlated error in the EKF prediction is corrected and the covariance is greatly reduced. This high reduction will be exploited for the recovery of high-innovation inliers: as correlations have weakened, consensus for the set will not be necessary to compute and individual compatibility will suffice to discard inliers from outliers.

An individual Gaussian prediction $\mathbf{h}^j \sim \mathcal{N} \left(\hat{\mathbf{h}}^j, \mathbf{S}^j \right)$ will be computed for each high innovation for every match $\mathbf{z}^j$ by propagating the state after the first partial update $\mathbf{x}_{k|k}$ through the projection model. The match will be accepted as an inlier if it lies within the 99% probability region of the predicted Gaussian for the measurement.

After testing all the high-innovation measurements a second partial update will be performed with all the points classified as inliers $\mathbf{z}^{hi\_inliers}$, following the usual EKF equations.

It is worth remarking here that splitting the EKF update does not have a noticeable effect on the computational cost. If n is the state size and m the measurement vector size, and in the usual SLAM case where the state is much bigger than the locally measured set $n >> m$, the main EKF cost is the covariance update which is $\mathcal{O}\left(mn^2\right)$. If the update is divided into two steps of measurement vector sizes m_1 and m_2 ($m = m_1 + m_2$), this covariance update cost stays almost the same. Some other minor costs grow, like the Jacobian computation which has to be done twice. But also some others are reduced, like the measurement covariance inversion which is $\mathcal{O}\left(m^3\right)$. Nevertheless, the effect of the latter two is negligible and for most EKF estimation cases the cost is dominated by the covariance update and remains approximately the same.

4.4 1-Point RANSAC Extended Kalman Filter from a Monocular Sequence Input

As previously stated, the proposed 1-point RANSAC and EKF combination is used in this chapter for the particular case of visual estimation from a monocular camera. In this section, the general method detailed in section 4.3 specializes to this specific application.

4.4.1 State Vector Definition

The state vector at step k is composed of a set of camera parameters $\mathbf{x}_{C_k}$ and map parameters $\mathbf{y}$.

$$\hat{\mathbf{x}}_k = \begin{pmatrix} \hat{\mathbf{x}}_{C_k} \\ \hat{\mathbf{x}}_M \end{pmatrix} ; \; \mathbf{P}_k = \begin{pmatrix} \mathbf{P}_{C_k} & \mathbf{P}_{C_k M} \\ \mathbf{P}_{M C_k} & \mathbf{P}_M \end{pmatrix} . \tag{4.9}$$

The estimated map $\mathbf{x}_M$ is composed of n point features $\mathbf{y}_i$; $\mathbf{x}_M = \left(\mathbf{y}_1^\top \ldots \mathbf{y}_n^\top\right)^\top$. Point features are parametrized in inverse depth coordinates $\mathbf{y}_{\rho,i} = \left(X_i \; Y_i \; Z_i \; \theta_i \; \phi_i \; \rho_i\right)^\top$ and converted to Euclidean parametrization $\mathbf{y}_{XYZ,i} = \left(X_i \; Y_i \; Z_i\right)^\top$ if and when the projection equation becomes linear enough, as described in chapter 3.

4.4.2 Dynamic Model

The dynamic model applied to the camera depends on the information available. For the case of pure visual estimation from a monocular sequence, a constant velocity model is sufficient for smooth hand-held motion. Details for the constant velocity model were already discussed in section 3.2.1. The camera state is then formed by position $\mathbf{r}_{C_k}$, orientation $\mathbf{q}_{C_k}$, and linear and angular velocities $\mathbf{v}$ and ω:

$$\mathbf{x}_{C_k} = \begin{pmatrix} \mathbf{r} \\ \mathbf{q} \\ \mathbf{v} \\ \omega \end{pmatrix} . \qquad (4.10)$$

When other sensorial information apart from the monocular sequence is available, it should be incorporated as input to the dynamic model. In this chapter, the combination of monocular vision plus wheel odometry is analyzed. In this case, the camera state only needs to contain position and orientation $\mathbf{x}_{C_k} = \begin{pmatrix} \mathbf{r} \\ \mathbf{q} \end{pmatrix}$. In the experiments shown at the end of this chapter, the classical model for a differential drive robot [Borenstein *et al.* 1996] has been chosen to model its dynamics.

4.4.3 Camera-Centered Estimation

It is well known that the usual EKF SLAM formulation, referred to a world reference frame, is only valid for local estimation in the surroundings of a sensor. Figure 4.3(a) illustrates the problem of this formulation: as the sensor moves away from the world reference, and if a pure exploratory trajectory is performed, the uncertainty of the estimation will always grow. Eventually it will reach a point where large linearization errors will cause inconsistency and filter divergence.

Figure 4.3(b) illustrates an alternative approach that alleviates this problem, that was first presented for EKF SLAM in [Castellanos *et al.* 2004]. It basically consists of referring all geometric parameters to a reference frame attached to the camera. Uncertainty in the locality of the sensor will always be kept low, greatly reducing the linearization errors associated with the measurement model. The camera-centered approach was first used for visual EKF estimation in [Civera *et al.* 2009c]; and has been thoroughly benchmarked in [Williams 2009].

The modifications with respect to world-centered visual SLAM are now given in detail. First, the state vector is composed of the location of the world reference frame $\mathbf{x}^{C_kW}$ and the map of estimated features $\mathbf{y}^{C_k}$, both expressed in the current camera reference frame C_k:

$$\mathbf{x}_k^{C_k} = \begin{pmatrix} \mathbf{x}^{C_kW} \\ \mathbf{y}^{C_k} \end{pmatrix} . \qquad (4.11)$$

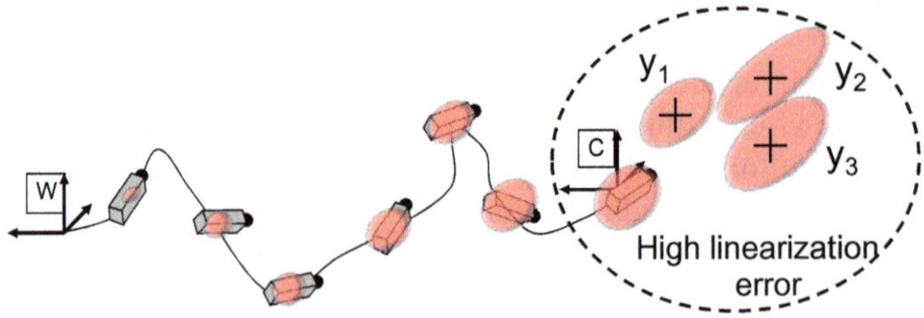

(a) World-referenced EKF estimation. As the camera moves away from the World reference and does not revisit known places the uncertainty grows both in the camera and newly initialized features y_i. The wide uncertainty regions will produce high linearization errors and filter inconsistency.

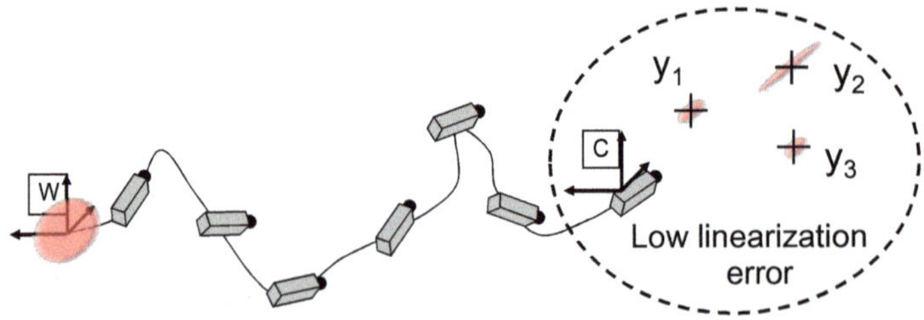

(b) Camera-centered EKF estimation. The camera location uncertainty is close to zero, as the reference is always attached to it. Uncertainties for features in its vicinity y_i will be also kept low, so measurement model linearization errors will be kept small for the whole estimation.

Fig. 4.3. Camera-centered and World-referenced EKF estimation.

The location of the world reference with respect to the current camera $\mathbf{x}^{C_kW} = \begin{pmatrix} \mathbf{r}^{C_kW} \\ \mathbf{q}^{C_kW} \end{pmatrix}$ is coded with its position $\mathbf{r}^{C_kW}$ and quaternion orientation $\mathbf{q}^{C_kW}$. When odometry information is not available and a constant velocity model is assumed, velocities should also be included in the state $\mathbf{x}_k^{C_k} = \begin{pmatrix} \mathbf{x}^{C_kW} \\ \mathbf{v}^{C_k} \\ \omega^{C_k} \\ \mathbf{y}^{C_k} \end{pmatrix}$.

For the prediction step at time k, the world reference frame and feature map are kept in the reference frame at time $k-1$ and a new feature $\mathbf{x}^{C_{k-1}C_k}$ that represents the motion of the sensor between $k-1$ and k is added:

$$\mathbf{x}_{k|k-1}^{C_{k-1}} = \begin{pmatrix} \mathbf{x}^{C_{k-1}W} \\ \mathbf{y}^{C_{k-1}} \\ \mathbf{x}^{C_{k-1}C_k} \end{pmatrix} \tag{4.12}$$

The predicted camera motion is represented in terms of position and orientation, represented via a quaternion:

$$\mathbf{x}^{C_{k-1}C_k} = \begin{pmatrix} \mathbf{r}^{C_{k-1}C_k} \\ \mathbf{q}^{C_{k-1}C_k} \end{pmatrix} . \tag{4.13}$$

The 1-point RANSAC EKF algorithm is applied with minor changes. The dynamic model of the system is applied over the motion relative to the previous frame contained in $\mathbf{x}^{C_{k-1}C_k}$, either using the constant velocity model in section 3.2.1 (in which case velocities should be kept then in the state as described above) or wheel odometry inputs. The pinhole camera model that serves as measurement model described in chapter 3 remains the same.

The algorithm proceeds then as explained in section 4.3. At the end of the algorithm, after the second update, a rigid transformation is applied to change the reference frame from the previous camera to the current one. The world reference location is updated:

$$\mathbf{r}^{C_kW} = \mathbf{R}^{C_kC_{k-1}} \left(\mathbf{r}^{C_{k-1}W} - \mathbf{r}^{C_{k-1}C_k} \right) \tag{4.14}$$

$$\mathbf{q}^{C_kW} = \mathbf{q}^{C_kC_{k-1}} \times \mathbf{q}^{C_{k-1}W} , \tag{4.15}$$

and the parameters representing motion from the previous to the current frame $\mathbf{x}^{C_{k-1}C_k}$ are marginalized out from the state. Inverse depth and Euclidean map features are also affected by this composition step:

$$\mathbf{y}_{\rho,i}^{C_k} = \begin{pmatrix} \mathbf{R}^{C_kC_{k-1}} \left(\begin{pmatrix} x_i^{C_{k-1}} \\ y_i^{C_{k-1}} \\ z_i^{C_{k-1}} \end{pmatrix} - \mathbf{r}^{C_{k-1}C_k} \right) \\ \mathbf{m}^{-1} \left(\mathbf{R}^{C_kC_{k-1}} \mathbf{m} \left(\theta_i^{C_{k-1}}, \phi_i^{C_{k-1}} \right) \right) \\ \rho_i \end{pmatrix} \tag{4.16}$$

$$\mathbf{y}_{XYZ,i}^{C_k} = \mathbf{R}^{C_kC_{k-1}} \left(\mathbf{y}_{XYZ,i}^{C_{k-1}} - \mathbf{r}^{C_{k-1}C_k} \right) . \tag{4.17}$$

In the previous formulas, $\mathbf{R}^{C_kC_{k-1}}$ stands for the rotation matrix from the camera reference C_k to C_{k-1}. This rotation matrix is computed from the quaternion $\mathbf{q}^{C_{k-1}C_k}$ in the state vector using the standard quaternion to rotation matrix conversion. The quaternion $\mathbf{q}^{C_kC_{k-1}}$ is the conjugate of the state vector one $\mathbf{q}^{C_{k-1}C_k}$. $\mathbf{m}$ stands for the transformation from azimuth-elevation angles to unit directional vector; and $\mathbf{m}^{-1}$ for the inverse transformation from a unit vector to its azimuth-elevation representation.

The covariance is updated using the Jacobians of this composition function $\mathbf{J}_{C_{k-1} \rightarrow C_k}$

$$\mathbf{P}_k^{C_k} = \mathbf{J}_{C_{k-1} \rightarrow C_k} \mathbf{P}_k^{C_{k-1}} \mathbf{J}_{C_{k-1} \rightarrow C_k}^\top . \tag{4.18}$$

4.5 Experimental Results

4.5.1 Benchmark Method for 6 DOF Camera Motion Estimation

The first step of the method takes an image sequence of the highest resolution, in order to achieve the highest accuracy. In this chapter, a 1224×1026 sequence was taken at 22 frames per second. A sparse subset of n camera locations $\mathbf{x}_{BA}^{C_1}$ are estimated by Levenberg-Marquardt Bundle Adjustment with robust likelihood model [Triggs *et al.* 2000] over the correspongding n images in the sequence $\{I_1, \ldots I_n\}$. Images are manually selected to ensure they form a strong network. The reference frame is attached to the camera C_1, corresponding to the first frame of the sequence I_1. For the experiments in sections 4.5.2 and 4.5.3, 62 overlapping camera locations were reconstructed by manually matching 74 points spread over the images. $15-20$ points are visible in each image.

$$\mathbf{x}_{BA}^{C_1} = \begin{pmatrix} \mathbf{x}_{1,BA}^{C_1} \\ \vdots \\ \mathbf{x}_{n,BA}^{C_1} \end{pmatrix}, \tag{4.19}$$

$$\mathbf{x}_{i,BA}^{C_1} = \left(X_{i,BA}^{C_1} \; Y_{i,BA}^{C_1} \; Z_{i,BA}^{C_1} \; \phi_{i,BA}^{C_1} \; \theta_{i,BA}^{C_1} \; \psi_{i,BA}^{C_1} \right)^\top . \tag{4.20}$$

Each camera location is represented by its position $\left(X_{i,BA}^{C_1} \; Y_{i,BA}^{C_1} \; Z_{i,BA}^{C_1} \right)^\top$ and Euler angles $\left(\phi_{i,BA}^{C_1} \; \theta_{i,BA}^{C_1} \; \psi_{i,BA}^{C_1} \right)^\top$. The covariance of the solution is computed by back-propagation of reprojection errors $P_{BA}^{C_1} = \left(J^\top R^{-1} J \right)^{-1}$, where J is the Jacobian of the projection model and R is the covariance of the Gaussian noise assumed in the model.

The input sequence is then reduced by dividing its width and height by four. The algorithm to benchmark is applied over the subsampled sequence. The reference frame is also attached to the first camera C_1, which is taken to be the same first one as in Bundle Adjustment. Images for which a Bundle Adjustment estimation is available are selected and stored $\mathbf{x}_{i,MS}^{C_1}$, each along with its individual covariance $P_{i,MS}^{C_1}$ directly extracted from the EKF at each step.

As the reference has been set to the same first image of the sequence, the Bundle Adjustment and sequential estimation solutions only differ in the scale of the reconstruction. So, in order to compare them, the relative scale s is estimated first by minimizing the error between the two trajectories. The Bundle Adjustment trajectory is then scaled $\mathbf{x}_{BA}^{C_1} = f_{scale}\left(\mathbf{x}_{BA}^{C_1} \right)$ and also its covariance $P_{BA}^{C_1} = J_{scale} P_{BA}^{C_1} J_{scale}^\top$.

Finally, the error is computed as the relative transformation between the two solutions:

$$e = \oplus \mathbf{x}_{BA}^{C_1} \ominus \mathbf{x}_{MS}^{C_1} ; \tag{4.21}$$

and the corresponding covariance of the error is computed by propagating the covariances of the global optimization and sequential estimate:

$$P_e = J_{eBA} P^{C_1}_{BA} J^{\top}_{eBA} + J_{eMS} P^{C_1}_{MS} J^{\top}_{eMS} \,. \tag{4.22}$$

It was checked in the experiments in the chapter that the covariance term from Bundle Adjustment, $J_{eBA} P^{C_1}_{BA} J^{\top}_{eBA}$, was negligible with respect to the summed covariance P_e. Since this is the case, it is our opinion that the Bundle Adjustment results can be considered as a reliable ground truth to evaluate sequential approaches. In the following figures, only uncertainty regions coming from filtering, $J_{eMS} P^{C_1}_{MS} J^{\top}_{eMS}$ are shown.

The same subsampled sequence was used for all the experiments in the following sections 4.5.2 and 4.5.3. The camera moves freely in six degrees of freedom in a computer lab, with the maximum distances between camera locations around 5 metres. Filter tuning parameters were equal for all the experiments: motion dynamic and measurement model noise were kept the same, the number of measured features in the image was limited to 30 and all the thresholds (e.g. for feature deletion, cross-correlation, inverse depth to Euclidean conversion and initialization) were also kept the same. The reader should be aware that despite all of care taken, the experiments are not exactly the same: One of the reasons is that the outlier rate is different for each method; some methods need to initialize more features in order to keep measuring 30. Nevertheless, in the opinion of the authors, this is the fairest comparison as the algorithms try always to measure always the same number of points and hence gather an equivalent amount of sensor data.

Figure 4.4 shows example images from the sequence used in the following two sections for 1-point RANSAC and JCBB benchmarking. The 62 camera locations from the 2796 images long sequence are also displayed. Results for different experiments using this benchmarking method have been grouped for better visualization and comparison: Figures 4.5 and 4.7 show estimation errors for different tunings of 1-point RANSAC and JCBB; and 4.9 details their computational cost. All the experiments in the chapter were run on an Intel(R) Core(TM) i7 processor at 2.67GHz.

4.5.2 1-Point RANSAC

First, the performance of 5-point and 1-point RANSAC is compared, in order to ensure that there is no degradation of performance when the sample size is reduced. Figures 4.5(a) and 4.5(b) show the errors of both algorithms with respect to the reference camera motion, along with their 99% uncertainty regions. It can be observed that reducing the sample size from 5 to 1 does not have a significant effect either on the accuracy or the consistency of the estimation. On the contrary, the figure even shows 1-point outperforming 5-point RANSAC. We attribute this to the fact that the theoretical number of hypotheses given by equation 4.1 was not inflated in our experiments, unlike in classical SfM algorithms [Raguram *et al.* 2008]. By increasing

(a) Sample images from the sequence used for benchmarking.

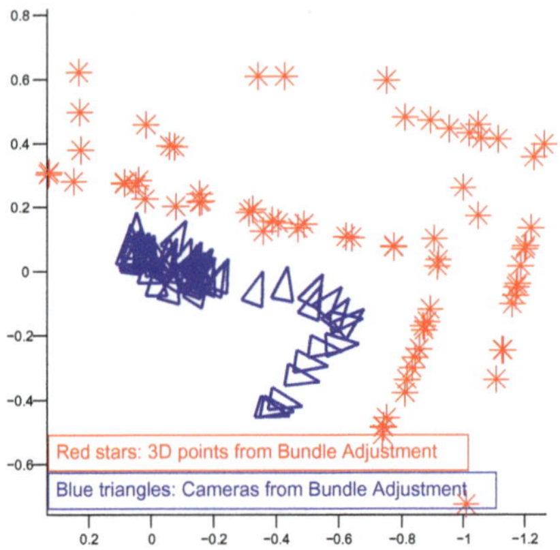

(b) Reference solution computed by the proposed method.

Fig. 4.4. Images extracted from the sequence used in the experiments and reference camera positions extracted.

the number of iterations, 5-point RANSAC results comes close to 1-point; but we find it remarkable that without this augmentation 1-point RANSAC already shows good behaviour. The standard deviation of image noise was chosen to be 0.5 for the experiments, as subpixel matching is used.

While the accuracy and consistency remains similar, the computational cost is much higher for the usual 5-point RANSAC than the proposed 1-point. The detail of the computational cost of both algorithms can be seen in Figures 4.9(a) and 4.9(b). The cost of RANSAC is low compared with the rest of the EKF computations for the 1-point case, but it is several orders of magnitude higher and is the main cost in the 5-point case. This is caused by the increase in the number of random hypotheses in frames with a large number of spurious matches. Figures 4.6(a) and 4.6(b) show the number of hypotheses in both cases, revealing that in 5-point RANSAC this is two orders of magnitude. The five higher green pikes appearing in all the figures are caused by dropped frames in the sequence where there is a jump in camera location. The correspondence search cost is increased at these frames, but notice that the cost of RANSAC is not increased at all.

Hypothesis generation from a single point opens the possibility of exhaustive rather than random hypotheses generation: while an exhaustive generation of all the possible combinations of 5 points in the measurement subset would be impractical, an exhaustive generation of 1-point hypotheses implies only as many hypotheses as measurements. Figure 4.5(c) details the errors for the 1-point exhaustive hypotheses generation case. Compared with 1-point random hypotheses generation in Figure 4.6(b), we observe similar accuracy and consistency. Figure 4.6(c) shows the number of iterations needed for comparison with the random adaptive case (Figure 4.6(b)). The computational cost is increased but, as shown in Figure 4.9(c), it is still

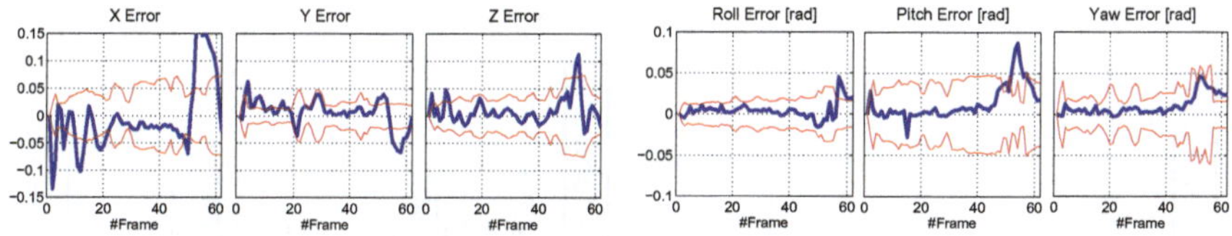

(a) 5-point RANSAC, $\sigma_z = 0.5$ pixels

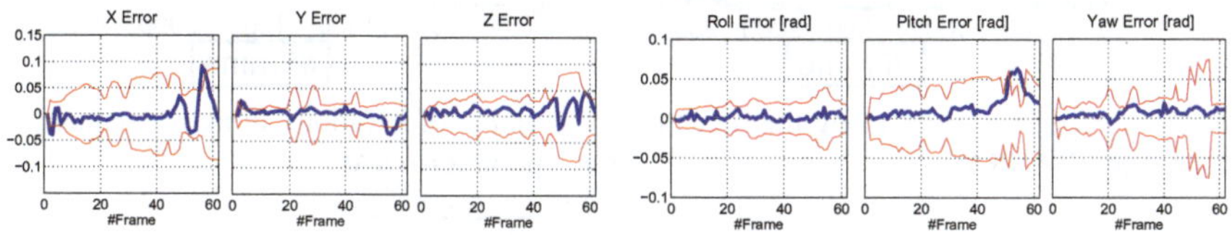

(b) 1-point RANSAC, $\sigma_z = 0.5$ pixels

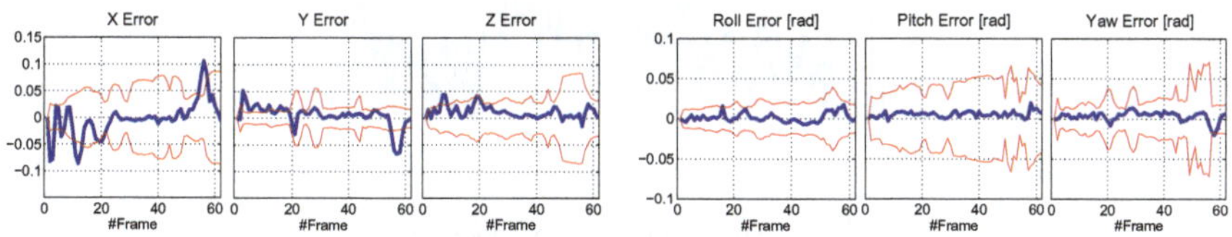

(c) 1-point exhaustive hypothesis, $\sigma_z = 0.5$ pixels

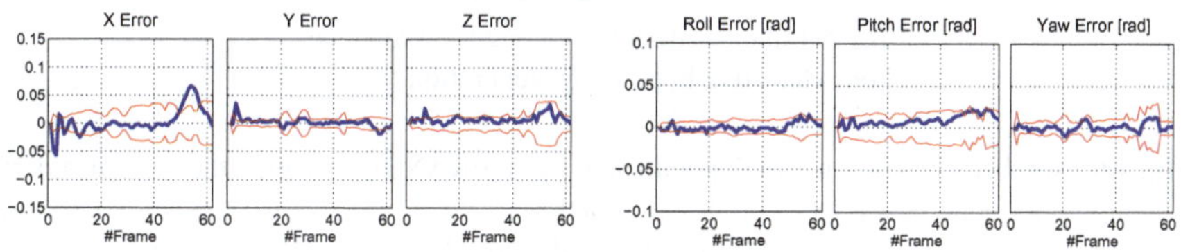

(d) 1-point RANSAC, $\sigma_z = 0.2$ pixels

Fig. 4.5. Camera location error (in thick blue) and uncertainty (in thin red) for different RANSAC configurations. Similar error and consistency is shown for 5-point and 1-point RANSAC in Figures 4.5(a) and 4.5(b) respectively. Figure 4.5(c) also reports similar results for exhaustive hypothesis testing. Figure 4.5(d) shows smaller errors as a result of making 1-point RANSAC stricter by reducing the standard deviation of measurement noise.

dominated by the EKF update cost. Both options are then suitable for real-time implementation, with the cheaper adaptive random 1-point RANSAC algorithm being preferable as performance is not degraded significantly.

From analyzing the computational cost in Figure 4.9(b) it can be concluded that the cost for 1-point RANSAC is always low compared with EKF computation even when the spurious match rate is high (the spurious match rate is shown in Figure 4.8(b)). As will be shown later, the latter becomes an important advantage over JCBB whose cost grows exponentially with the rate of spurious matches. This efficiency opens the possibility of making the RANSAC algorithm stricter by reducing the measurement noise standard deviation and hence discarding high noise points

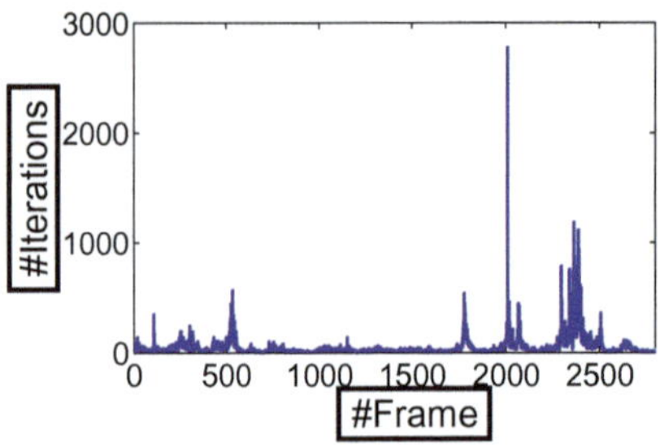

(a) Number of iterations along the sequence for 5-point RANSAC.

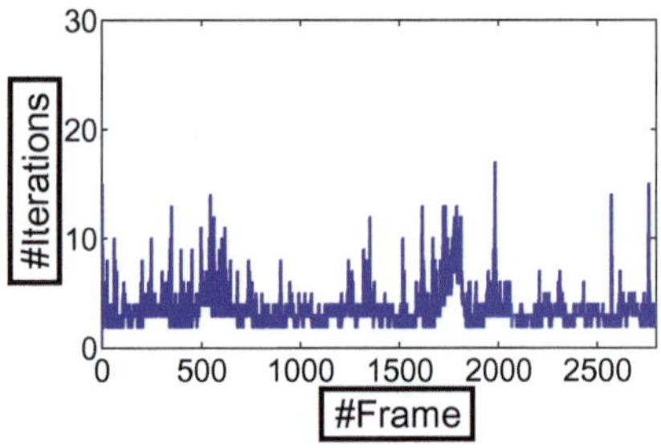

(b) Number of iterations along the sequence for 1-point RANSAC.

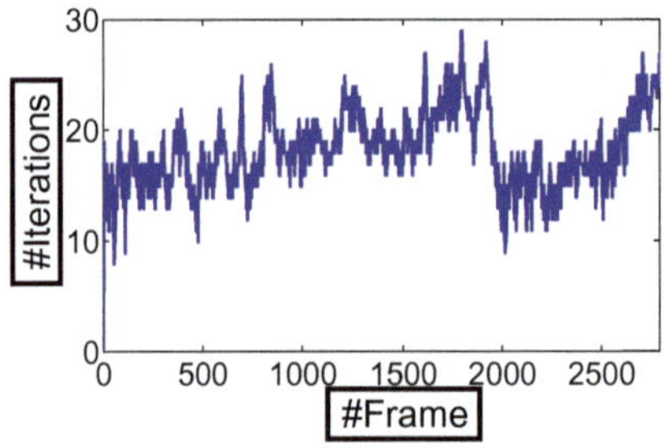

(c) Number of iterations along the sequence for exhaustive hypotheses generation.

Fig. 4.6. Number of iterations for 5-points and 1-point RANSAC. Notice the several orders of magnitude increase for the 5-point case, causing a large cost overhead when compared with 1-point RANSAC (Figures 4.9(a), 4.9(b) and 4.9(c) detail the computational cost for the three cases respectively).

in the EKF. Such analysis can be done by reducing the standard deviation from 0.5 to 0.2 pixels: high noise points were discarded as outliers, as can be seen in Figures 4.8(b) and 4.8(d). The computational cost increases, as shown in Figure 4.9(e), but still remains small enough to reach real-time performance at 22 Hz. The benefit of discarding high noise points can be observed in Figure 4.5(d): errors and their uncertainty were reduced (but still kept mostly consistent) as a result of measuring more accurate points.

4.5.3 Joint Compatibility Branch and Bound (JCBB)

RANSAC and JCBB tuning is a thorny issue when benchmarking both algorithms. As both cases assume Gaussian distributions for the measurement and decide based on probability, we considered it fairest to choose equal significance levels for the probabilistic tests of both algorithms. The significance level was chosen to be 0.05

in the χ^2 test that JCBB performs to ensure joint compatibility for the matches. Consistently, the probabilistic threshold for RANSAC was set to 95% for voting (line 15 in the algorithm in section 4.3) and for the rescue of high-innovation matches (line 29 in the algorithm in section 4.3).

The results of benchmarking JCBB are shown in the following figures. First, figure 4.7(a) details the errors and uncertainty regions for the EKF using JCBB. It can be observed that the estimation in Figure 4.7(a) show larger errors and inconsistency than the 1-point RANSAC one in Figure 4.7(b), repeated here for visualization purposes. The reason can be observed in Figure 4.8, where the outlier rates for 1-point RANSAC and JCBB are shown: the number of matches considered outliers by 1-point RANSAC is greater than by JCBB. The points accepted as inliers by JCBB are the ones that spoil the estimation.

A stricter version of JCBB has been benchmarked by reducing the standard deviation of uncorrelated measurement noise to 0.2 pixels, as was done with 1-point RANSAC. The spurious match rate for both algorithms, shown in Figure 4.8(c) and 4.8(d), shows that 1-point RANSAC remains more discriminative and hence produces more accurate estimation than JCBB (Figure 4.7(c)). 1-point RANSAC errors for the same tuning are repeated in 4.7(d) for comparison purposes. Also, as previously noted, the computational cost of JCBB grows exponentially when made stricter: Figure 4.9(f) shows peaks over a second in the worst cases.

JCBB can also be made stricter by increasing the significance level α of the χ^2 test it performs to check the joint compatibility of the data. Several experiments were run varying this parameter. The lowest estimation errors, shown in Figure 4.7(e), were reached for $\alpha = 0.5$ instead of the usual $\alpha = 0.05$. Estimation errors for this best JCBB tuning are still larger than in any of the 1-point RANSAC experiments.

4.5.4 Trajectory Benchmarking against GPS

The following sections benchmark the presented filtering scheme for the estimation of long camera trajectories. The benchmarking method of the previous section becomes difficult to apply here, so camera translation only is benchmarked against GPS data. This section describes the benchmarking procedure.

Similarly to the previous section, our EKF estimation takes the first camera frame C_1 as the frame of reference. A similarity transformation (rotation $\mathbf{R}^{WC_1}$, translation $\mathbf{t}^{WC_1}$ and scale s) has to be applied which aligns every point of the trajectory $\mathbf{r}^{C_1C_k} = \left[x^{C_1C_k}\ y^{C_1C_k}\ z^{C_1C_k}\right]^\top$ with the GPS data $\mathbf{r}^{WGPS_k}$, whose frame of reference we will denote by W:

$$\begin{bmatrix} \mathbf{r}^{WC_k} \\ 1 \end{bmatrix} = \begin{bmatrix} x^{WC_k} \\ y^{WC_k} \\ z^{WC_k} \\ 1 \end{bmatrix} = \begin{bmatrix} s\mathbf{R}^{WC_1} & \mathbf{t}^{WC_1} \\ \mathbf{0} & 1 \end{bmatrix} \begin{bmatrix} x^{C_1C_k} \\ y^{C_1C_k} \\ z^{C_1C_k} \\ 1 \end{bmatrix} . \tag{4.23}$$

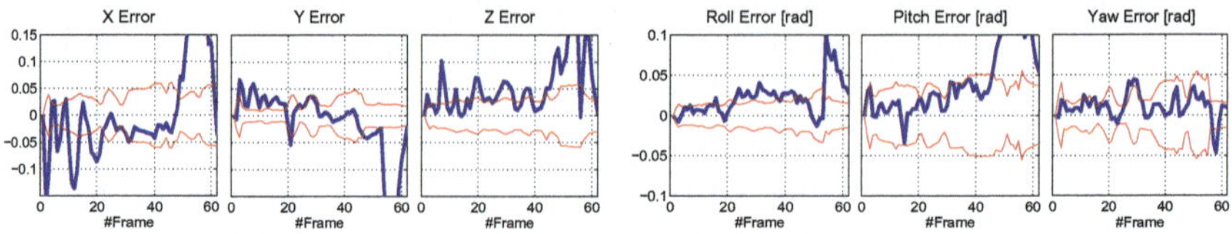

(a) JCBB, $\sigma_z = 0.5$ pixels

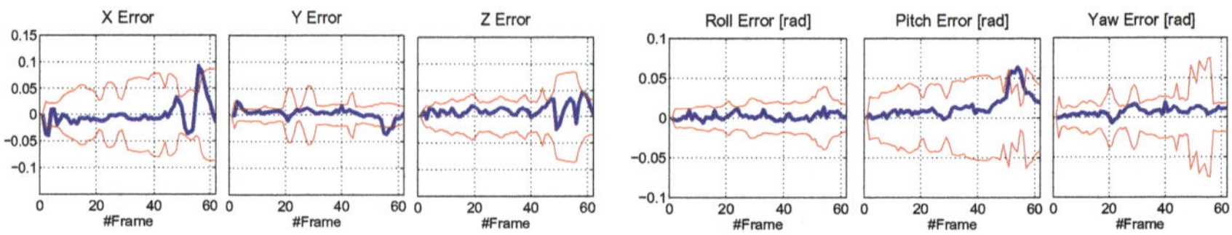

(b) 1-point RANSAC, $\sigma_z = 0.5$ pixels

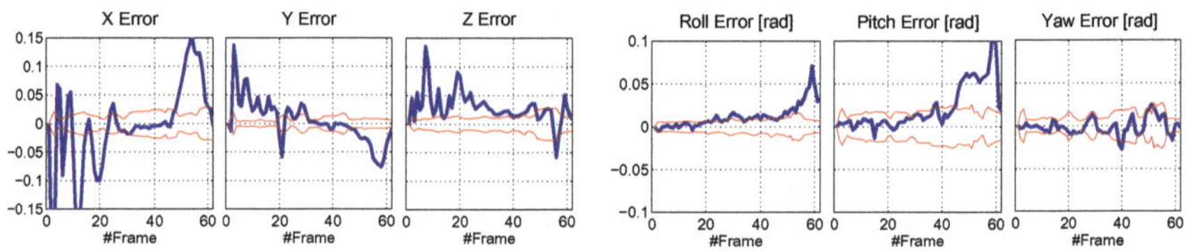

(c) JCBB, $\sigma_z = 0.2$ pixels

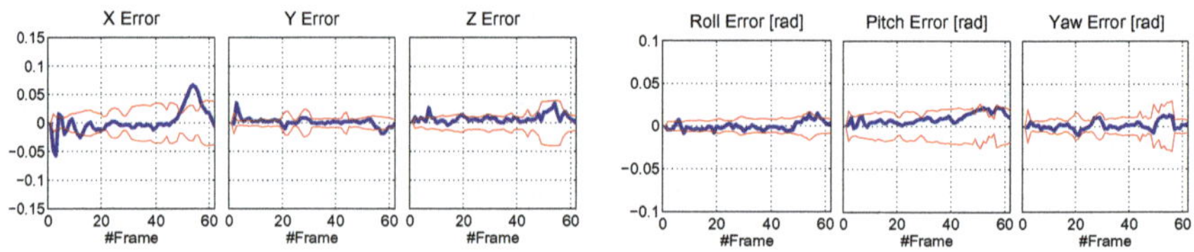

(d) 1-point RANSAC, $\sigma_z = 0.2$ pixels

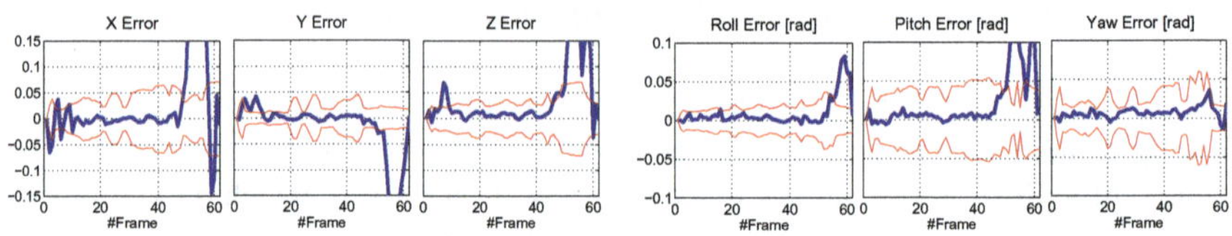

(e) JCBB, $\sigma_z = 0.2$ pixels, $\alpha = 0.5$

Fig. 4.7. Camera location errors when using JCBB is shown in Figures 4.7(a) and 4.7(c), for standard deviations of 0.5 and 0.2 pixels respectively. Figures 4.7(b) and 4.7(d) showing 1-point RANSAC results for the same filter tuning are repeated here for comparison. It can be seen that 1-point RANSAC outperforms JCBB in both cases. Figure 4.7(e) shows the best JCBB tuning found by the authors, which still gives worse results than 1-point RANSAC.

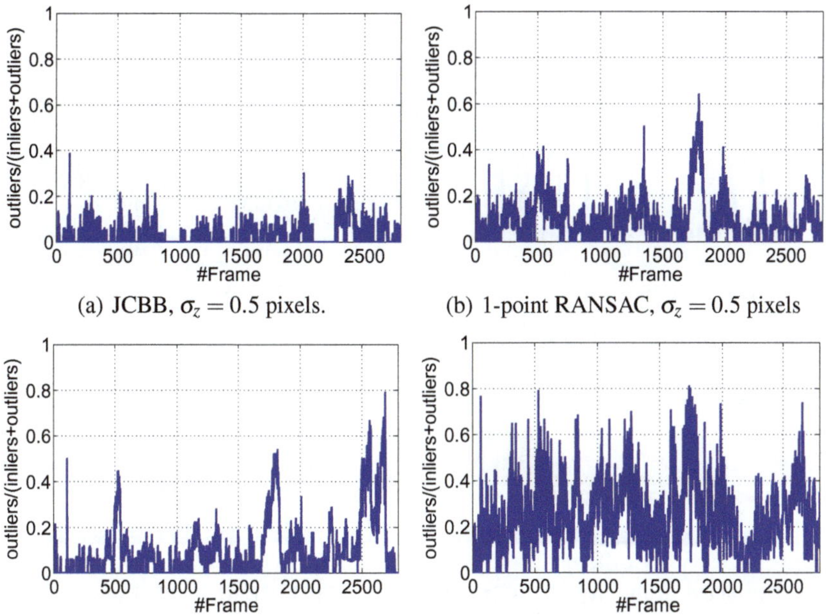

(a) JCBB, $\sigma_z = 0.5$ pixels.

(b) 1-point RANSAC, $\sigma_z = 0.5$ pixels

(c) JCBB, $\sigma_z = 0.2$ pixels

(d) 1-point RANSAC, $\sigma_z = 0.2$ pixels.

Fig. 4.8. Spurious match rate for JCBB and RANSAC when measurement noise standard deviation σ_z is reduced to 0.2 pixels. It can be observed that reducing the measurement noise makes both techniques stricter, but 1-point RANSAC remains more discriminative.

The value of $\mathbf{t}^{WC_1}$ is taken from the GPS data in the first camera frame. Trajectory estimation from pure monocular vision will not be able to recover the scale s, which will remain unknown. For the combination of a monocular camera and wheel odometry input, the overall scale of the estimation is observed by odometry readings and then $s = 1$ in Equation 4.23. The rotation between GPS and the first camera position $\mathbf{R}^{WC_1}$ will be unknown in every case, as it is non-observable from GPS readings.

The unknown parameters of the alignment (s and $\mathbf{R}^{WC_1}$ for pure monocular, and only $\mathbf{R}^{WC_1}$ for monocular plus wheel odometry) are obtained via a non-linear optimization that minimizes the error between the aligned trajectory $\mathbf{r}^{WC_k}$ and the GPS trajectory $\mathbf{r}^{WGPS_k}$.

For the sake of simplicity, the assumption that the position of the camera sensor and the GPS antenna coincide on the robot has been made in the above reasoning, which is reasonable as the position of the sensors differ by only a few centimetres and robot paths cover hundreds of metres.

Finally, the error of each camera position in the reconstructed path is computed as the Euclidean distance between each point of the estimated camera path and GPS path, both in the W reference:

$$e_k = \sqrt{\left(\mathbf{r}^{WC_k} - \mathbf{r}^{WGPS_k}\right)^\top \left(\mathbf{r}^{WC_k} - \mathbf{r}^{WGPS_k}\right)}. \quad (4.24)$$

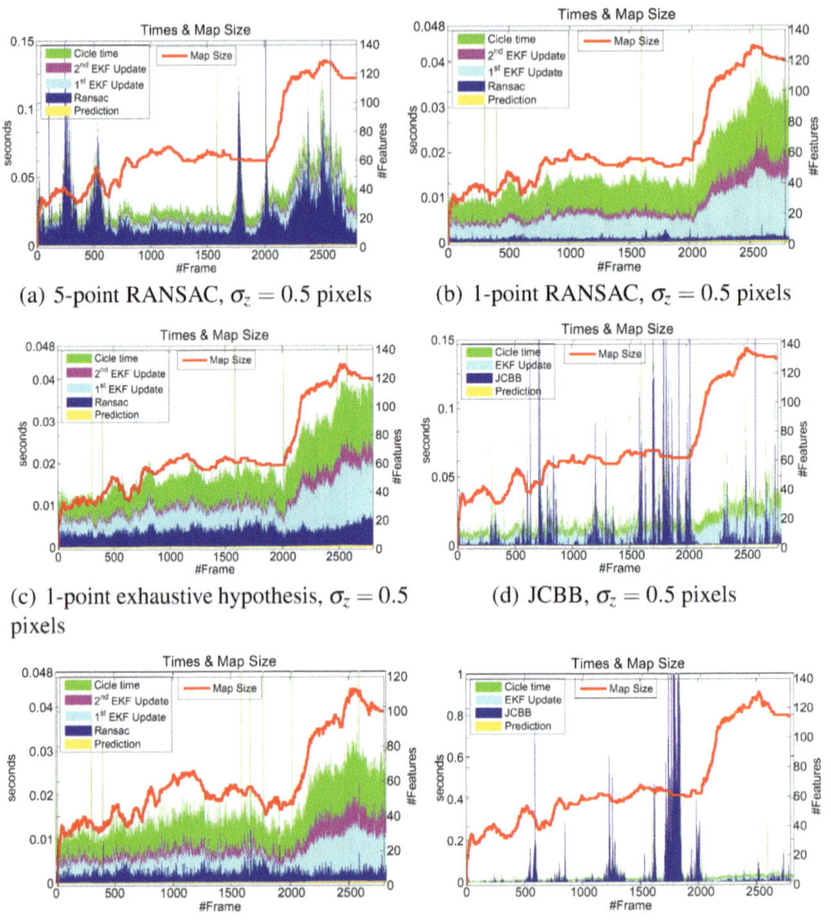

(a) 5-point RANSAC, $\sigma_z = 0.5$ pixels

(b) 1-point RANSAC, $\sigma_z = 0.5$ pixels

(c) 1-point exhaustive hypothesis, $\sigma_z = 0.5$ pixels

(d) JCBB, $\sigma_z = 0.5$ pixels

(e) 1-point RANSAC, $\sigma_z = 0.2$ pixels

(f) JCBB, $\sigma_z = 0.2$ pixels

Fig. 4.9. Detail of times and map sizes for different RANSAC and JCBB configurations in double y-axis figures: times are shown as areas and measured in seconds on the left y-axis; the map size is displayed as a a red line and is measured on the right y-axis. 1-point RANSAC exhibits much lower computational cost than 5-point RANSAC and JCBB. 1-point RANSAC also shows only a small increase when made exhaustive or stricter, making it suitable for real-time implementation at 22 Hz for the map size detailed in the figures.

4.5.5 Pure Monocular EKF-Based Estimation for Long Sequences

Three different sequences from the *RAWSEEDS* dataset have been used to test the validity of the 1-point RANSAC EKF for long-term camera motion estimation. All sequences were recorded by a 320×240 Unibrain camera with a wide-angle lens capturing at 30 fps.

In the first sequence, consisting of 6000 images, the robot translates for about 146 metres. The second sequence has 5400 images and the robot describes a similar trajectory length, about 153 metres. Finally, a very long and challenging sequence is evaluated that consists of 24180 frames (13.5 minutes of video) in which the robot describes a trajectory of 650 metres. In this latter sequence, although the accumulated drift makes the error noticeable when plotted with the GPS trajectory, the relative error with respect to the trajectory keeps the same low value as the other two shorter sequences (1% of the trajectory length).

Figure 4.10 shows an image from the 650 metres experiment, along with the tracked features. It can be observed that around 1–2 hundred features per frame had to be measured in order to reduce scale drift error. This high number will increase the computational cost of the EKF beyond real-time bounds for the pure monocular case. In the particular experiments presented, the algorithm runs at about 1 Hz. Nevertheless, it will be shown in next subsection how introducing extra information about the scale will reduce the number of measurements, enabling real-time performance for the combination of visual tracking plus wheel odometry.

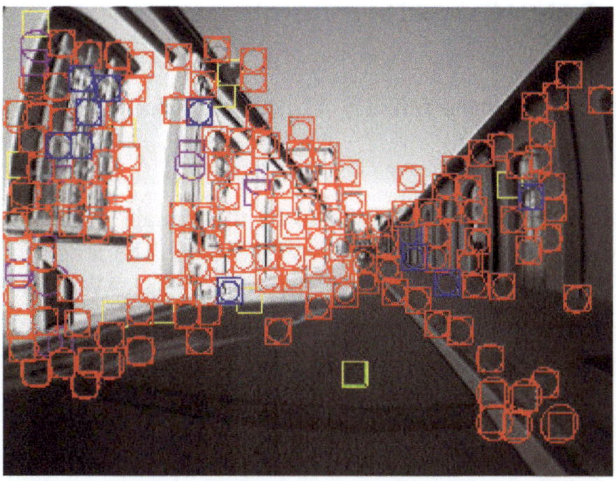

Fig. 4.10. Image from the 650 metres sequence, showing the high number of tracked features.

Figure 4.11 shows the estimated (in black) and the GPS (in red) trajectories over a top view extracted from Google Maps for each one of the sequences. The accuracy of the estimated trajectories is clear from visual inspection. Table 4.1 details the maximum and mean errors obtained in these experiments and also for the experiment in the next section combining monocular vision and wheel odometry inputs. Figure 4.12 shows histograms of the errors for the three sequences.

Table 4.1. EKF-based visual estimation error for long camera trajectories.

Trajectory length [m]	Sensor used	Mean error [m]	Maximum error [m]	% mean error over the trajectory
146	monocular	1.3	4.2	0.9%
153	monocular	1.9	3.3	1.1%
650	monocular	6.4	11.1	1.0%
1310	monocular and wheel odometry	9.8	23.6	0.7%

Subfigures 4.12(c) and 4.12(d) in this latter figure show histograms of the errors for the 650 metres experiment in two different versions of the 1-point RANSAC algorithm: the first one of them using the algorithm 1 and the second one replacing the random hypotheses generation with exhaustive hypotheses generation as evaluated in Figure 4.5(c). The conclusion from section 4.5.2 is confirmed here: exhaustive hypothesis generation only very slightly improves the estimation errors; so adaptive random 1-point RANSAC should be preferred.

4.5.6 *Visual Odometry from a Monocular Sequence Plus Wheel Odometry*

Figure 4.13 shows the trajectory obtained by the visual odometry algorithm over a GoogleMaps plot and compared against GPS data. The length of the estimated trajectory is about 1310 metres and was covered by the RAWSEEDS mobile robot in 30 minutes, capturing 54000 frames. The maximum and mean error were 23.6 and 9.8 metres respectively. Adding wheel odometry information allowed us to reduce the number of tracked features to 25, enabling real-time operation at 30 frames per second.

The processing time per frame for this sequence using 1-point RANSAC can be observed in Figure 4.14 in the form of a histogram. It can be noticed that the total computational cost per step is under 33 milliseconds in 98% of the frames, suggesting that the algorithm is suitable for real-time implementation. It can be observed in the right-hand figure that for the same number of image measurements JCBB's computational cost far exceeds real-time constraints in a large number of

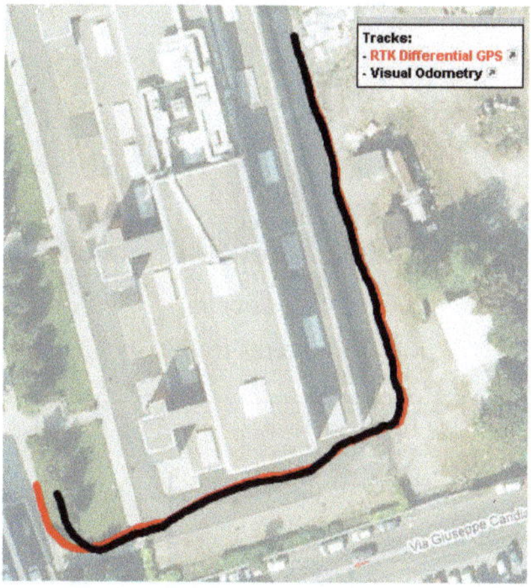

(a) 146 metre trajectory

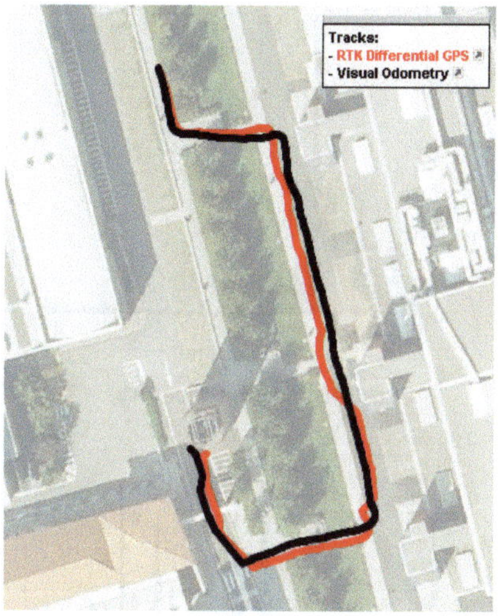

(b) 156 metre trajectory

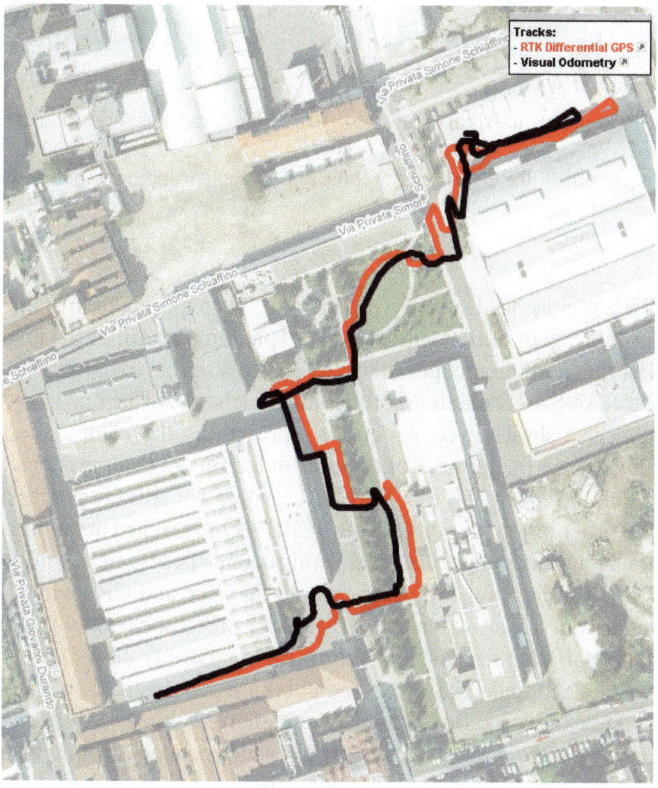

(c) 650 metre trajectory

Fig. 4.11. Estimated trajectories from pure monocular data and GPS data.

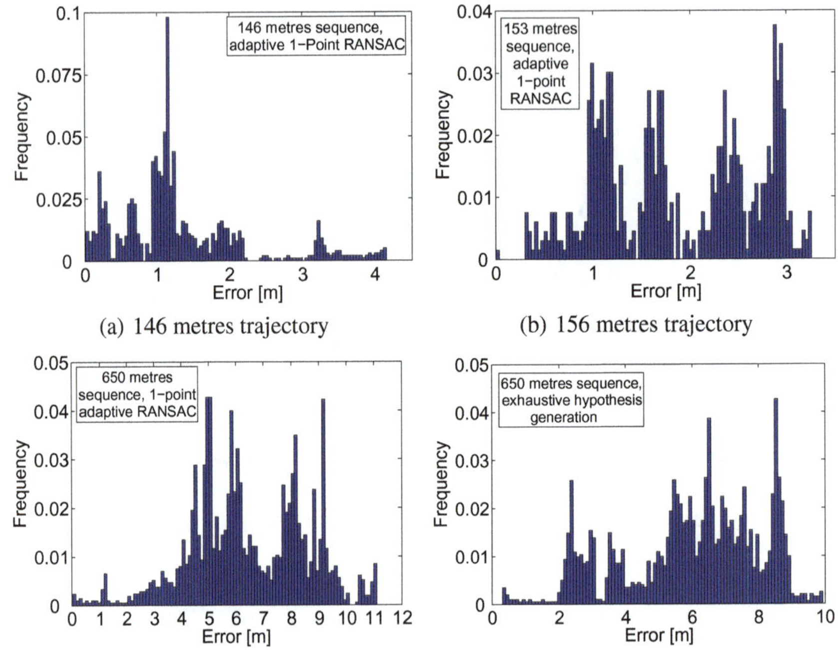

(a) 146 metres trajectory

(b) 156 metres trajectory

(c) 650 metres trajectory

(d) 650 metres trajectory; Exhaustive-SAC

Fig. 4.12. Histograms of the errors for the three experiments using only monocular information.

frames. JCBB's exponential complexity arises in this experiment in frames where a significant proportion of outliers are present, expanding the tail of the histograms of the figure. For this particular experiment, JCBB's histogram expands to 2.4 seconds while 1-Point RANSAC's maximum time only reaches 0.44 seconds.

Figure 4.14 also shows two histograms representing the computational cost of both algorithms when the number of features in the image is increased to 50. It can be observed that the cost of 1-Point RANSAC grows, but still the processing cost is always on the order of tenths of a second. JCBB's cost reaches maximum values of several hours, and processing times of several seconds per frame are not unusual

Figure 4.15(b) shows raw odometry as a red thin line and GPS with a blue thick line for comparison. It can be observed that early drift appears and the plotted trajectory is rather far from the GPS locations. Figure 4.15(a) shows pure monocular estimation in thin red and GPS measurements in thick green. Observing this plot carefully, it can be observed that a monocular camera is able to very accurately estimate orientation, but the unobservability of the scale produces drift in this parameter for the number of tracked features (25) considered in this experiment.

Finally, Figure 4.15(c) details the estimated trajectory that can be achieved from the combination of the two sensors. Accurate estimation is achieved for a trajectory

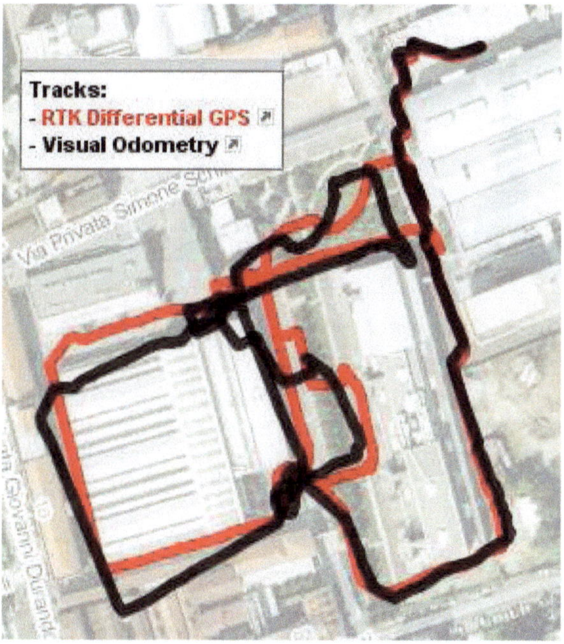

Fig. 4.13. Visual odometry results compared against RTK GPS over a Google Maps plot.

of 1.3 kilometres, which can be compared with state of the art in monocular visual odometry, e. g. [Scaramuzza *et al.* 2009].

4.6 Discussion

A novel RANSAC algorithm is presented in this chapter which, for the first time and differently from standard purely data-driven RANSAC, incorporates *a priori* probabilistic information into the hypothesis generation stage. As a consequence of using this prior information, the sample size for the hypothesis generation loop can be reduced to the minimum size of 1 point data. 1-point RANSAC has two main strengths worth summing up here. First, as in standard RANSAC, model constraints are checked *after* hypothesis data has been fused with the a priori model, an advantage over JCBB. Second, using 1-point plus prior knowledge hypotheses greatly reduces the number of hypotheses to construct and hence the computational cost compared with usual RANSAC based solely on data. Its linear cost in the state size also outperforms JCBB's exponential complexity in the number of outliers. In a practical sense, its linear complexity means an overhead of less than 10% of the standard EKF cost, making it suitable for real-time implementation in local visual SLAM or SfM.

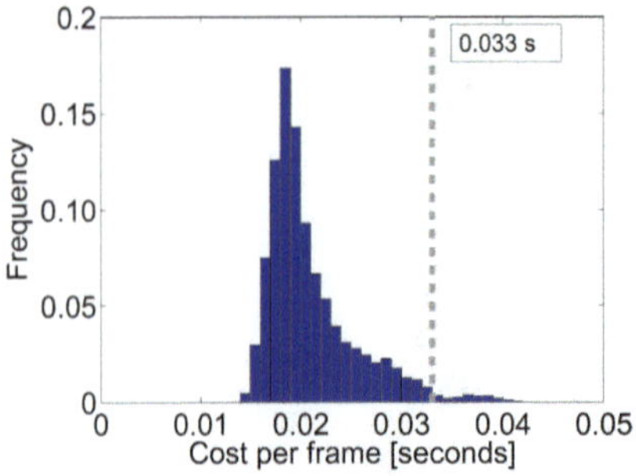

(a) 1-point RANSAC; 25 measured features per frame

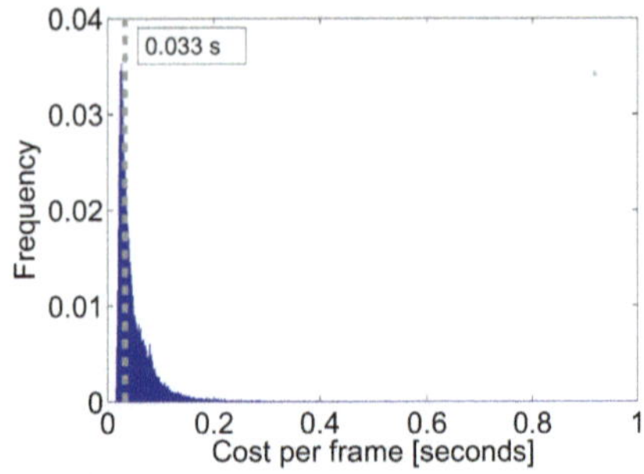

(b) JCBB; 25 measured features per frame

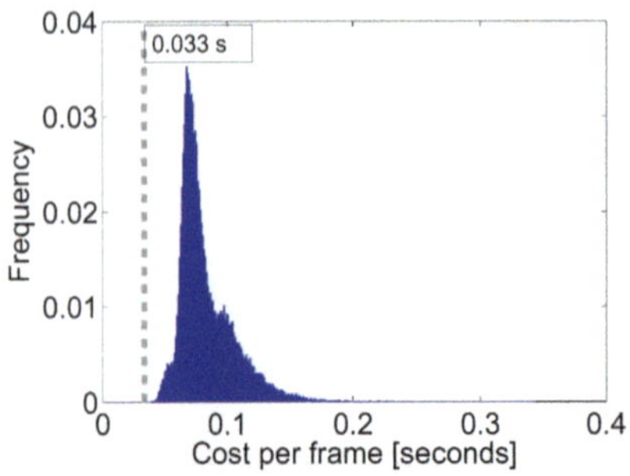

(c) 1-point RANSAC; 50 measured features per frame

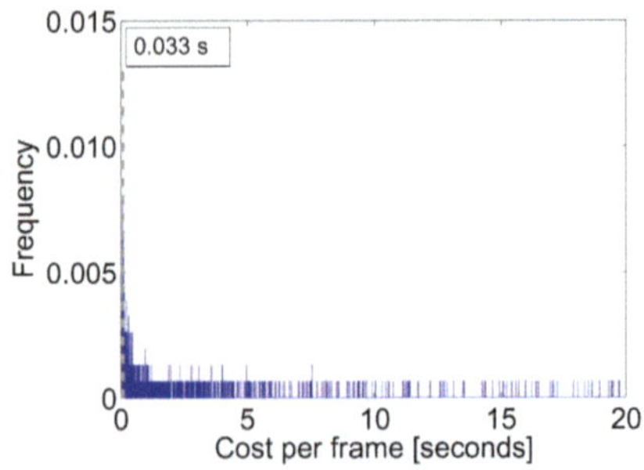

(d) JCBB; 50 measured features per frame

Fig. 4.14. Histograms showing the computational cost for RANSAC and JCBB for the cases of 25 and 50 image points per frame. Experiment 4.14(d) had to be early terminated at frame 1533, as JCBB computational cost rises in some frames up to 1544 seconds

While the relevance of algorithms like JCBB or the recent Active Matching (AM) reside on their generality, the main advantage in the presented approach is its efficiency. 1-point RANSAC is directed to the particular case of a rigid scene. The rich variety of correlation patterns that a covariance matrix can encode is manageable by general methods like JCBB or AM. Our 1-point RANSAC exploits the very simple pattern where all the correlations are mainly explained by sensor motion, and hence small size data subsets are enough to constraint the rest of the measurements. For more complex models, like non-rigid scenes or multi-object tracking, 1-point RANSAC may not offer such a satisfactory result.

Nevertheless, it is also true that estimation from a moving sensor's data stream in an almost rigid scene covers a great percentage of SLAM problems; and a specific method more efficient than general methods can be of importance. In this sense, 1-point RANSAC outperforms existing approaches by presenting lower cost and scaling well with the state vector and measurement size, and also with the outlier rate.

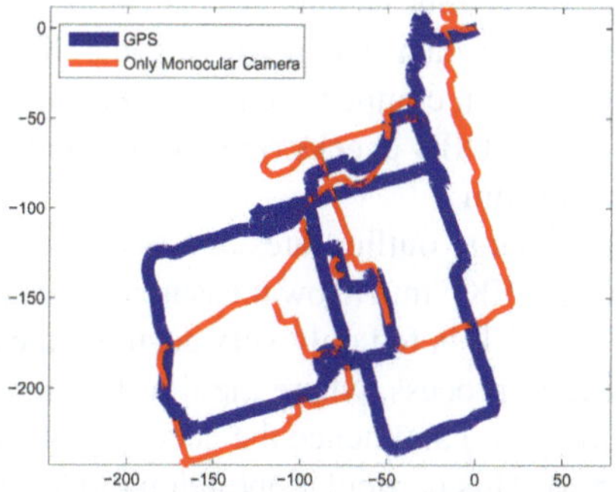

(a) Pure monocular estimation (thin red) tracking 25 features and GPS trajectory (thick blue). Large errors appear caused by scale drift, which is unobservable by a monocular camera.

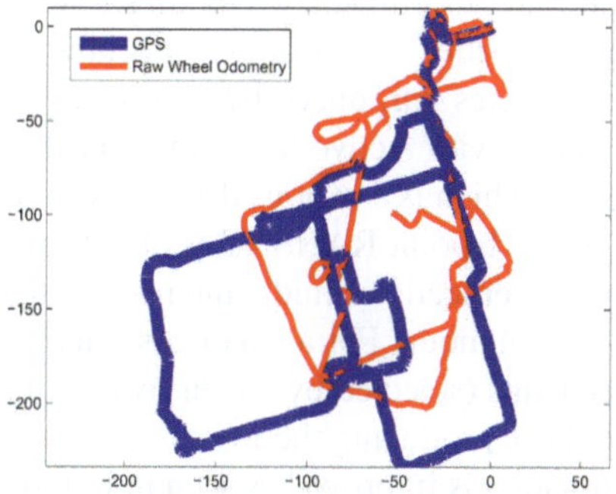

(b) Raw odometry measurements (thin red) and GPS trajectory (thick blue). Errors in raw odometry are caused by early drift typical from proprioceptive sensors.

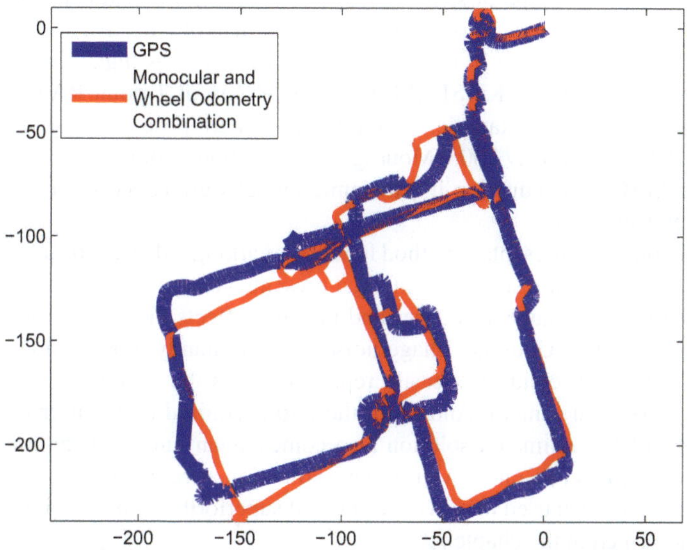

(c) Visual Odometry estimation from the combination of monocular camera plus wheel odometry (thin red) and GPS trajectory (thick blue). The combination of both sensors overcomes their deficiencies when used alone. Real-time performance at 30 Hz can be achieved, and error is 0.7% of the trajectory.

Fig. 4.15. Pure monocular estimation showing scale drift in Figure 4.15(a), raw odometry input showing drift in Figure 4.15(b) and visual odometry results combining the two in Figure 4.15(c); all are compared against GPS trajectory (thick blue line).

Besides its efficiency, 1-point RANSAC also has some advantages in dealing with non-linearities as a result of checking rigidity after data fusion where some of the inaccuracies introduced by non-linearities have been compensated. This advantage is shared with Active Matching. On the contrary JCBB checks rigidity before data fusion, which is a serious drawback of the algorithm.

Since 1-point RANSAC is able to deal with large outlier rates at low computational overhead, we find it interesting to force the EKF into a low measurement error operation mode. For a small cost increase, the EKF is fed only very accurate measurements (selected by "a survival of the fittest" process, where the fittest features are those producing the lowest error measurements) and hence the accuracy of the estimation is improved as seen in Figure 4.5(d). This particular operation mode can only be achieved due to the efficiency of the presented algorithm, being impractical if spurious match rejection is expensive.

It is also worth remarking that although this book is focused on the particular case of EKF monocular SLAM, the new 1-point RANSAC method presented here is independent of the type of sensor used. The only requirement is the availability of highly correlated prior information, which is typical of EKF SLAM for any kind of sensor used — and also in the multisensor case. Also, as highly correlated priors are not exclusive to EKF SLAM, the applicability of 1-point RANSAC could be even broader. As an example, we think that camera pose tracking in keyframe schemes [Klein & Murray 2008, Mouragnon *et al.* 2009] would benefit from our 1-point RANSAC cost reduction if a dynamic model were added to predict camera motion between frames.

This chapter also presents a method for benchmarking 6 degrees of freedom camera motion estimation results. The method presents three clear advantages: First, it is intended for real image sequences and includes effects difficult to reproduce by simulation (like non-Gaussian image noise, shaking handy motion, image blur or complex scenes). Second, it is easily reproducible as the only hardware required is a high resolution camera. And third, the effort required by the user is low. The uncertainty of the estimated solution also comes as an output of the method and the appropriateness of Bundle Adjustment estimation as reference can be validated. The method has been used to prove the claimed superiority of the 1-point RANSAC method described in the chapter.

The general EKF plus 1-point RANSAC algorithm has been experimentally tested for the case of large camera trajectories in outdoor scenarios. Sensor-centered filtering instead of the traditional world-centered method has been used in order to reduce the uncertainty in the area local to the current camera and reduce linearization errors. For the pure monocular case, errors around 1% of the trajectory have been obtained for trajectories up to 650 metres from a publicly available dataset. The number of tracked features in the image has to be increased to 1–2 hundreds in order to avoid scale drift. This high number makes this case currently moves us away from real-time performance, and the method runs at 1 frame per second.

The combination of monocular vision and wheel odometry has also been benchmarked for the visual odometry application. The extra odometric information makes scale observable; the number of tracked features can be reduced and real-time performance can be achieved for this case. A 1300 metre long trajectory has been estimated in the chapter, with the mean error against GPS coming out at 0.7% of the trajectory.

Chapter 5
Degenerate Camera Motions and Model Selection

Abstract. The assumption of a general camera motion –translation and rotation– between frames in an image sequence leads to inconsistent estimations when the camera performs more restricted motions, like pure rotation, or even no motion. In this situation, the noise present in the data fits the extra terms in the overparametrized model and artificially estimates dimensions for which we have no information. This chapter presents an Interacting Multiple Models (IMM) framework which can switch automatically between parameter sets in monocular SLAM; selecting the most appropriate motion model at each step. Remarkably, this approach of full sequential probability propagation means that there is no need for penalty terms to achieve the Occam property of favouring simpler models –this arises automatically. We demonstrate our method with results on a complex real image sequence with varied motion.

5.1 Introduction

Image sequence processing relies on camera motion models to actively identify potential point matches. Off-line methods rely on geometrical models relating two or three images to robustly compute matches. In [Torr *et al.* 1999] it is shown how different models should be used at different parts of a general sequence to avoid degenerate geometries. This geometrical model selection has been extended to segment different motion models between image pairs or triplets [Schindler & Suter 2006, Kanatani 2004, Torr 2002].

In contrast to these two or three-view geometrical models, the probabilistic motion models used in SLAM are well suited to modelling long sequences of close images instead of discrete sets of images. However a single probabilistic model can similarly only deal with sequences which follow the prescribed model or processing will fail. In this chapter, the monocular EKF SLAM is extended to deal with more

J. Civera et al.: Structure from Motion Using the Extended Kalman Filter, STAR 75, pp. 99–110.
springerlink.com

than one probabilistic motion model, expanding the range of sequences compatible with the priors represented by a set of tuning parameters. We use a sequential Bayesian approach to model selection.

Thanks to Bayesian probability propagation, monocular SLAM with a general translating camera can deal with low parallax motions — such as rotations — provided that the camera re-observes map features whose locations are well-estimated as a result of parallax observed previously in the sequence, and so model switching is not a must in some cases where it would be in the off-line approaches. However, when monocular SLAM is initialised on-the-fly without a known scene pattern, model selection is an issue. If the camera initially undergoes a low parallax motion, no reliable estimation is possible. Any measurement noise may be considered parallax by the filter producing inconsistent depth estimates. We tackle this problem with model selection.

Multiple model methods are well known in maneuvering target tracking. An excellent and recent survey of this can be found in [Li & Jilkov 2005]. In this chapter, we adapt to the SLAM problem the most widespread of those methods, Interacting Multiple Models (IMM), initially proposed by Blom in [Blom & Bar-Shalom 1988]. The IMM estimator is a suboptimal hybrid filter — that is, it estimates the continuous values of a proccess, and the discrete probabilities of a set of models — whose main features are: 1) It assumes that the system can jump between the members of a set of models, which is the case of our monocular SLAM estimation, and 2) It offers the best compromise between complexity and performance.

Thanks to the use of multiple models, the range of image sequences that can be processed with a single system tuning is enlarged. We work with a bank of 7 models: one model of a stationary camera, three models of pure rotation motion (constant angular velocity) with different angular acceleration covariances, and three general translation + rotation models (constant velocity, constant angular velocity) with different angular and linear acceleration covariances. Via the Bayesian model selection of IMM, the system prefers simpler (less general) models where they fit the data. As a result, the search regions for the predicted image features are smaller than with a single model EKF. These reduced search regions increase mismatch rejection and reduce the processing cost of image search. Additionally, the computed probabilities per model allow the segmentation of a sequence into different models.

Section 5.2 discusses and formulates the sequential Bayesian model selection. The Interacting Multiple Model approach to Bayesian model selection is detailed in 5.3. Some details about the use of IMM in the SLAM problem are given in section 5.4. Section 5.5 verifies the method using real imagery and shows how it deals with sequence bootstrap. Finally section 5.6 summarises the paper's conclusions.

5.2 Bayesian Model Selection for Sequences

In standard single-model monocular SLAM algorithms, Bayes' rule combines at every step past estimation information with current image data. Given the background information I and the image data at current step D, the posterior probability density function for the set of parameters θ defining our model M is updated via Bayes' formula:

$$p(\theta|DMI) = p(\theta|MI)\frac{p(D|\theta MI)}{p(D|MI)} \ . \tag{5.1}$$

In this chapter we consider cases where a single model M is not sufficient to cover all of the sequences we would like to track. Taking full advantage of the fully probabilistic estimation that our SLAM approach is performing, we formulate our multiple model problem in a Bayesian framework.

Consider, as Jaynes does in Chapter 20 of his book [Jaynes 2003], a discrete set of models $\mathscr{M} = \{M^1, \ldots, M^r\}$ — rather than a single one — which might feasibly describe the assumptions of a sequential SfM process. We start by assigning initial scalar probabilities $P(M^1|I), \ldots, P(M^r|I)$ which represent prior belief about the different models based on background information I, and which are normalised to add up to one. If no prior information exists, these probabilities may well be assigned initially equal.

At each new image, where we acquire image measurements data D, we update the probability of each model according to Bayes' rule:

$$P(M^j|DI) = P(M^j|I)\frac{P(D|M^jI)}{P(D|I)} \tag{5.2}$$

In this expression, the first term is the probability of the model being correct given only the prior information. In the fraction, the numerator is the likelihood of obtaining the data given that the model is correct. The denominator is the normalizing constant, computation of which can be avoided when the posterior probabilities of a mutually-exclusive set of models are all computed, or alternatively cancels out when the ratio of posterior probabilities of different models is calculated.

So, what is the likelihood $P(D|M^jI)$ of the data given a model in a monocular SLAM system? It is simply the joint likelihood of all of the feature measurements in an image:

$$P(D|MI) = \frac{1}{\sqrt{\left|2\pi\mathbf{S}_{k|k-1}\right|}}\exp\left(-\frac{1}{2}\nu^\top\mathbf{S}_{k|k-1}^{-1}\nu\right) , \tag{5.3}$$

where

$$\nu = \mathbf{z} - \mathbf{h}(\hat{\mathbf{x}}_{k|k-1}) , \tag{5.4}$$

$$\mathbf{S}_{k|k-1} = \mathbf{H}_{k|k-1}\left(\mathbf{F}\mathbf{P}_{k-1|k-1}\mathbf{F}^\top + \mathbf{G}\mathbf{Q}\mathbf{G}^\top\right)\mathbf{H}_{k|k-1}^\top + \mathbf{R} \ . \tag{5.5}$$

$\mathbf{F} = \frac{\partial \mathbf{f}}{\partial \mathbf{x}}$ and $\mathtt{G} = \frac{\partial \mathbf{f}}{\partial \mathbf{n}}$ are the Jacobians of the dynamic model $\mathbf{f}$ with respect to the state vector $\mathbf{x}$ and the process noise $\mathbf{n}$ respectively. $\mathbf{H}_{k|k-1} = \frac{\partial \mathbf{h}}{\partial \mathbf{x}_{k|k-1}}$ is the Jacobian of the measurement equation $\mathbf{h}$ with respect to the state vector $\mathbf{x}_{k|k-1}$. $\mathbf{S}_{k|k-1}$ is the covariance of the predicted image measurements, which is computed by adding the linear propagation the state vector uncertainty $\mathbf{P}_{k|k-1}$ and the image noise covariance $\mathbf{R}$.

We should note, as Jaynes explains with great clarity, that in this correctly formulated Bayesian approach to model selection there is no need for ad-hoc terms like Minimum Description Length which penalise 'complex' models and favour simple ones. The 'Occam principle' of selecting the simplest model which is able to capture the detail of the data and avoiding overfitting is taken care of automatically by correctly normalising our comparison of different models. The big difference between our approach and the common two-view model selection methods (e.g. [Kanatani 2004, Torr 2002, Schindler & Suter 2006]) which require penalty terms is that our concept of a model is probabilistic at its core, not just geometric (like homography, affine, ...). For our use in sequential probabilistic tracking, a model must actually define a probability distribution during a transition. This is what makes it possible to calculate proper likelihoods for the models themselves, independent of parameters.

The formulation above allows us to obtain posterior probabilities for our models in one frame, but we are interested in propagating these probabilities through a sequence. This is achieved by defining a vector of model probabilities — a 'state vector' for models or set of mixing weights:

$$\mu_{\mathbf{k|k}} = \left(\mu^{\mathbf{1}}_{\mathbf{k|k}} \cdots \mu^{\mathbf{r}}_{\mathbf{k|k}}\right)^{\top} . \tag{5.6}$$

We fill $\mu_{\mathbf{k|k}}$ with the prior model probabilities $P(M^1|I), \ldots, P(M^r|I)$ before processing the first image, and after processing use the values of $\mu_{\mathbf{k|k}}$ as the priors for each model in Equation 5.2 and then replace these values with the posterior values calculated.

A final step is needed in between processing images, which is to apply a mixing operator to account for possible transitions between models. With a homogeneous Markov assumption that the probability of transition from one model to any other is constant at any inter-frame interval, this is achieved by:

$$\mu_{\mathbf{k|k-1}} = \pi \mu_{\mathbf{k-1|k-1}} , \tag{5.7}$$

where π is a square matrix of transition probabilities where each row must be normalised. In the typical case that the dominant tendancy is sustained periods of motion with one model, this matrix will have large terms on the diagonal. If the models are ordered with some sense of proximity, the matrix will tend to have large values close to the diagonal and small ones far away.

The sequential process of calculating model probabilities therefore evolves as a loop of mixing and update steps and at motion transitions in the sequence evidence will accrue over several frames.

5.3 Interacting Multiple Model

IMM is presented in the tracking literature as a hybrid estimation scheme, well suited to estimating the continuous state of a system that can switch between several behaviour modes. This hybrid system is then composed of a continuous part (the state) and a discrete part (the behaviour modes). The continuous part of such a system is defined by its state and measurement equations:

$$\dot{\mathbf{x}}(t) = \mathbf{f}(\mathbf{x}(t), M(t), \mathbf{w}(t), t) \tag{5.8}$$

$$\mathbf{z}(t) = \mathbf{h}(\mathbf{x}(t), M(t), \mathbf{v}(t), t) \tag{5.9}$$

where the dynamics of the process and the measurements depend not only on the state $\mathbf{x}(t)$ and the process and measurement noise $\mathbf{w}(t)$ and $\mathbf{v}(t)$ at time t, but also on the model $M(t)$ that governs the system at time t. The probability of each of those

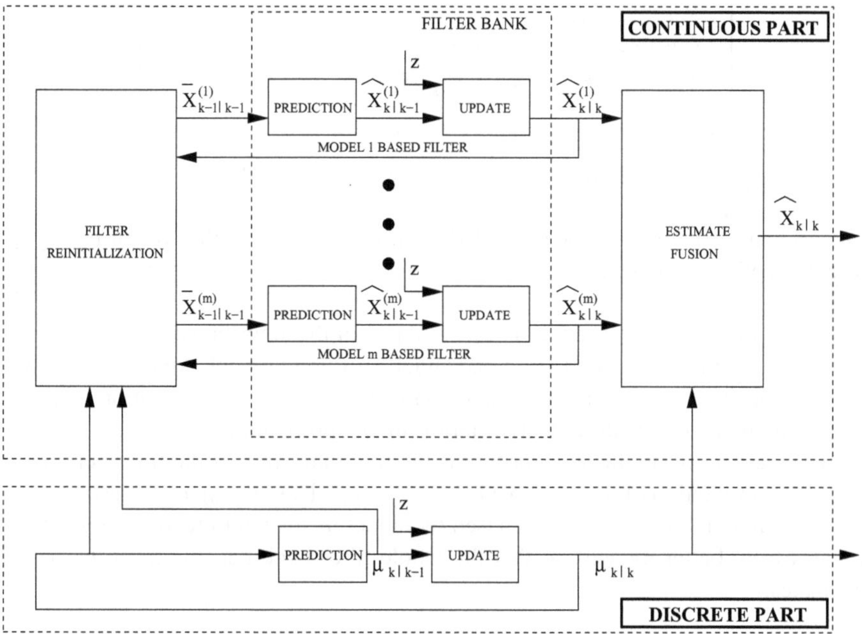

Fig. 5.1. Interacting Multiple Model algorithm scheme

1. Filter reinitialization (for $i = 1, 2, \ldots, r$):

`Predicted model probability:`

$\mu^i_{k|k-1} = P\{M^i_k | z^{k-1}\} = \sum_j \pi_{ji} \mu^j_{k-1}$

`Mixing weight:`

$\mu^{j|i}_{k-1} = P\{M^j_{k-1} | M^i_k, z^{k-1}\} = \pi_{ji} \mu^j_{k-1} / \mu^i_{k|k-1}$

`Mixing estimate:`

$\bar{\mathbf{x}}^i_{k-1|k-1} = E[\mathbf{x}_{k-1} | m^i_k, z^{k-1}] = \sum_j \mathbf{x}^j_{k-1|k-1} \mu^{j|i}_{k-1}$

`Mixing covariance:`

$$\bar{\mathbf{P}}^i_{k-1|k-1} = \sum_j (\mathbf{P}^j_{k-1|k-1} + \\ + (\bar{\mathbf{x}}^i_{k-1|k-1} - \hat{\mathbf{x}}^j_{k-1|k-1})(\bar{\mathbf{x}}^i_{k-1|k-1} - \hat{\mathbf{x}}^j_{k-1|k-1})^\top) \mu^{j|i}_{k-1}$$

2. EKF bank filtering (for $i = 1, 2, \ldots, r$):

`Prediction:` $\hat{\mathbf{x}}^i_{k|k-1}, \mathbf{P}^i_{k|k-1}, \mathbf{h}(\mathbf{x}^i_{k|k-1}), \mathbf{S}^i_{k|k-1}$

`Measurement:` $\mathbf{z}_k$

`Update:` $\hat{\mathbf{x}}^i_{k|k}, \mathbf{P}^i_{k|k}$

3. Model probability update (for $i = 1, 2, \ldots, r$):

`Model likelihood:` $L^i_k = \mathcal{N}(v^i_k; 0, S^i_k)$

`Model probability:` $\mu^i_k = \frac{\mu^i_{k|k-1} L^i_k}{\sum_j \mu^j_{k|k-1} L^j_k}$

4. Estimate fusion

`Overall state:`

$\hat{\mathbf{x}}_{k|k} = \sum_i \hat{\mathbf{x}}^i_{k|k} \mu^i_k$

`Overall covariance:`

$$\mathbf{P}_{k|k} = \sum_i \left(\mathbf{P}^i_{k|k} + (\hat{\mathbf{x}}_{k|k} - \hat{\mathbf{x}}^i_{k|k})(\hat{\mathbf{x}}_{k|k} - \hat{\mathbf{x}}^i_{k|k})^\top \right) \mu^i_k$$

Fig. 5.2. Interacting Multiple Model algorithm

models being effective at time t is coded in the discrete probability vector $\mu_{\mathbf{k-1|k-1}}$, as explained in section 5.2.

Figure 5.1 shows graphically the structure of the IMM estimator. The whole algorithm is detailed in Figure 5.2. The central part of the algorithm consists of a bank of r filters running in parallel, each one under a different model. An overall estimation for the state can be obtained as a sum of the a posteriori estimation of every filter weighted with the discrete a posteriori model probabilities.

A key aspect of the IMM algorithm is the reinitialisation of the filter before the parallel computation of the filter bank at every step. This mixing of the estimations allows individual poor estimates caused by model mismatch to recombine with estimates from better models, so that the whole filter bank benefits from the better estimates.

5.4 Interacting Multiple Model Monocular SLAM

Given the tracking-oriented IMM algorithm, some aspects have to be taken into account before applying it to our particular monocular SLAM problem.

1. **Active search and 1-point RANSAC:** In the multiple model tracking literature, little attention is given to the matching (data association) proccess, which is crucial in SLAM algorithms. If matching is mentioned, as in [Kirubarajan *et al.* 1998], it is said that the most general model, that is, the model with the largest covariance, is used to compute the measurement covariance for gating correspondences—the implication is 'always to expect the worst'.
   In monocular SLAM, most of the time this weakest search region is unnecesary large, increasing both the computational cost and the risk of obtaining a false match. A more realistic search region can be defined by the combination of the individual filters weighted by their discrete probabilities. The only assumption that has to be made is that motion changes are smooth, a reasonable assumption when dealing with image sequences at high frame rate. The form of the image search regions is therefore determined by the following equations:

$$\hat{\mathbf{h}}_{k|k-1} = \sum_i \hat{\mathbf{h}}^i_{k|k-1} \mu^i_{k|k-1} \tag{5.10}$$

$$\mathbf{S}_{k|k-1} = \sum_i (\mathbf{S}^i_{k|k-1} + (\hat{\mathbf{h}}_{k|k-1} - \hat{\mathbf{h}}^i_{k|k-1}) \tag{5.11}$$

$$(\hat{\mathbf{h}}_{k|k-1} - \hat{\mathbf{h}}^i_{k|k-1})^\top) \mu^i_{k|k-1} \tag{5.12}$$

   Spurious rejection is not mentioned in the multiple model literature either. In our monocular SLAM case, the 1-point RANSAC developed in the chapter 4 can be easily adapted to the Interacting Multiple Model framework.
2. **Map management:** As detailed in section 2.4.3, map management strategies for deleting bad features and adding new ones are convenient in monocular SLAM. We are also using inverse depth to cartesian conversion in order to reduce the computational cost of the algorithm as detailed in section 3.6.

5.5 Experimental Results

A 1374 frame sequence was recorded with a 320×240 wide-angle camera at 30fps. The camera makes a motion consisting of the following sequence of essential movements: stationary $\rightarrow$ pure rotation $\rightarrow$ general motion (translation and rotation) $\rightarrow$ pure rotation $\rightarrow$ stationary. The sequence has been processed using the dimensionless inverse depth formulation in the appendix B and two different types of motion modelling. Firstly, IMM EKF formulation with a bank of seven models: stationary camera, rotating camera (three angular acceleration levels with standard deviation 0.1, 0.5 and 1 pixels), and general motion (with 3 acceleration levels for both linear and angular components with standard deviations of 0.1, 0.5 and 1 pixels). Secondly, as a base reference, a single model for general motion with acceleration noise standard deviation of 1 pixel, both angular and linear. Both formulations are fed the same starting image feature detections. On analysing the results the advantages of the IMM over single model monocular SLAM become clear.

5.5.1 Consistent Start Up Even with Rotation

As was said in section 5.1, single model EKF SLAM leads to inconsistent mapping if the camera initially undergoes low parallax motion. In the analysed sequence, we have an extreme case of this as the camera is either stationary or rotating for more than 600 frames. Figure 5.3 compares the estimation results with a single model EKF and our IMM algorithm at step 600, when the camera has performed non-translational motion. Features are plotted as arrows if (as should be the case) no finite depth has been estimated after the no parallax motion. It can be observed that, for the single model case, all features have collapsed to narrow, *false*, depth estimates while in the IMM case all of the features have no depth estimation.

5.5.2 Low Risk of Spurious Matches due to Small Search Regions

It can be noticed in Figure 5.4 that although high process noise models are necessary in order to retain tracking features during high accelerations, these models are scarcely used for any length of time. In hand-held camera sequences, constant velocity motions are much more common than accelerated ones. This is reflected by the model probabilities, as we see that the highest probabilities are given to the lower acceleration noise models on most frames.

When using a single model estimation, we are forced to choose the most general model in order to maintain tracking under high acceleration. As process noise directly influences search region size, we are forced to maintain large search regions, unnecessary most of the time. As a consequence, the risk of obtaining false matches grows. As IMM selects at any time the most probable motion model, preferring simpler models, it adjust the search region to the real motion at any time, resulting in considerably reduced ellipses and lowering the risk of mismatches.

In Figure 5.5 the large factor of reduction in the size of search ellipses can be observed. Subfigure (a) shows a detail of a feature search region at frame 100, at the top using IMM and at the bottom using a single model. Search regions in subfigure (b) correspond to the same feature at frame 656, when camera starts translating and high acceleration is detected. Notice that the IMM ellipse slightly enlarges in adapting to this motion change, but continues to be smaller than the single-model one. Finally, (c) exhibits the consequences of having unnecessary big search regions: false correspondences happen. Due to mismatches like this one, the estimation in this experiment fails catastrophically.

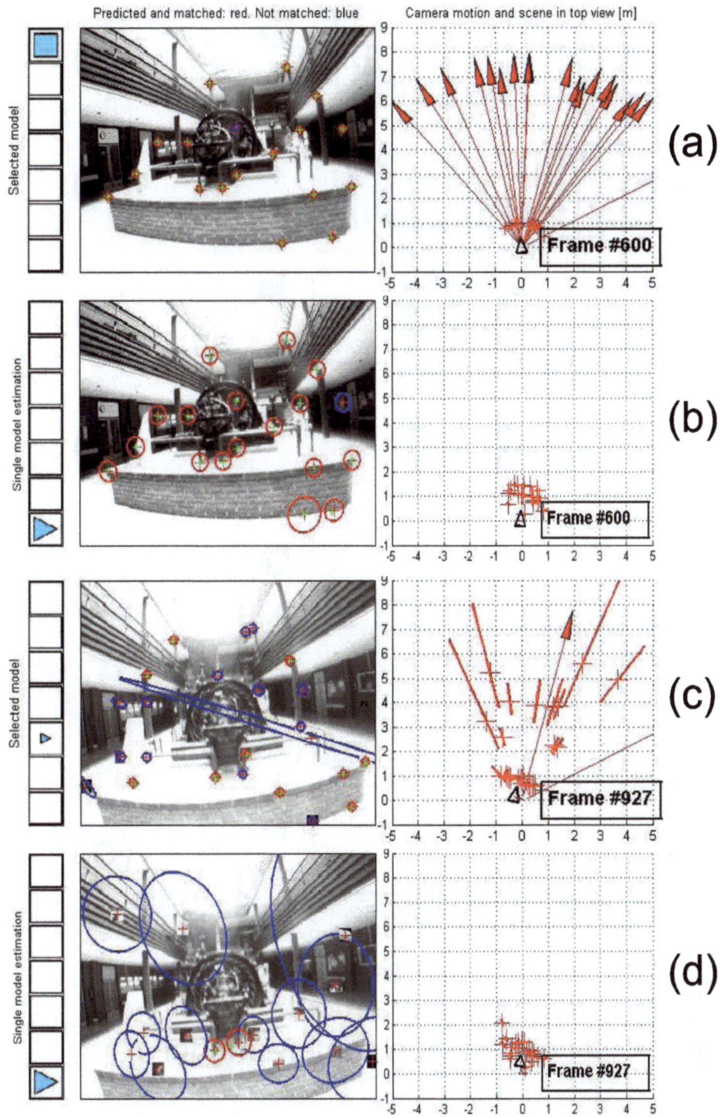

Fig. 5.3. (a, left) frame 600 and (a, right) 3D top view of the IMM estimation at this frame. The camera has been either stationary or rotating until this frame. It can be seen in Figure 5.4 that rotation and still camera models have high probability throughout this early part of the sequence. IMM, correctly, has not estimated any feature depth –features whose depths have not been estimated (their depth uncertainties, stored in inverse depth formulation, encompass infinity) are plotted as arrows–. (b), frame 600 and top-viewed estimation with single-model monocular SLAM. The overparametrized model has led to narrow, false depth estimates. When the camera translates this inconsistent map leads to false matches that cause the estimation to fail, as seen in (d) at frame 927 of the sequence. On the other hand, (c) shows the correct map estimation performed by the IMM algorithm at this same frame 927.

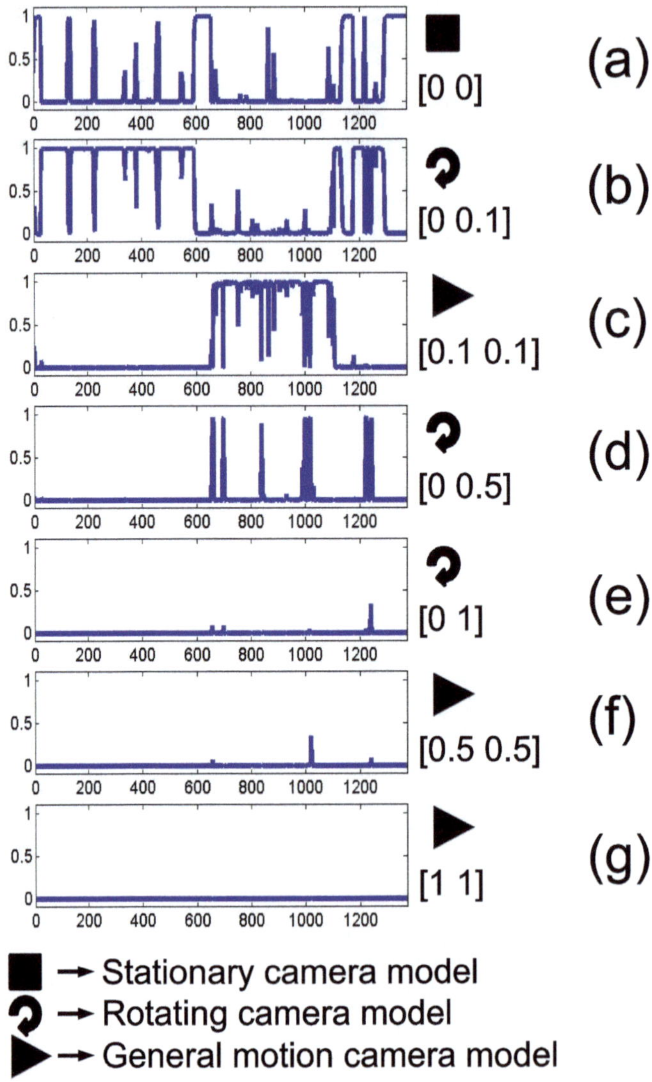

Fig. 5.4. Posterior model probabilities along the sequence. Each model is represented by its acceleration noise standard deviation$[\sigma_a, \sigma_\alpha]$ expressed in pixels, following the notation in [Civera *et al.* 2007b]. Notice that the probability for the most general model ($\sigma_a = 1pxl, \sigma_\alpha = 1pxl$) is always under 0.01. The stationary camera model (a) and low acceleration noise models (b) and (c) are assigned the highest probabilities in most of the frames. In spite of being rarely selected, the high acceleration noise models are important to keep the features track at the frames where motion change occurs (small spikes are visible at these points).

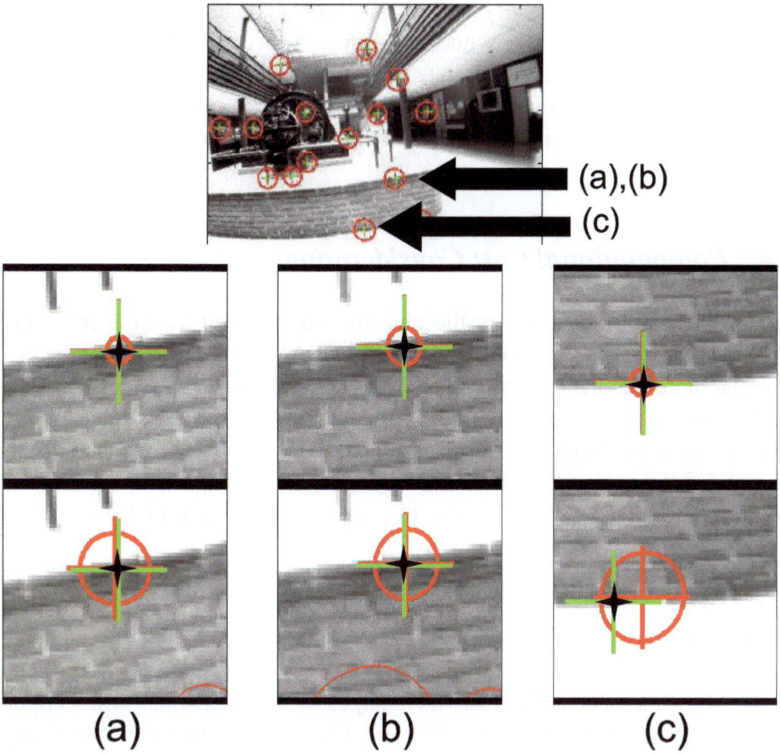

Fig. 5.5. (a), IMM (top) and single-model (bottom) feature search ellipse when the camera is rotating. (b), the same feature IMM and single-model search regions when the camera begins to translate. (c), mismatch in the single-model case caused by an unnecesary large ellipse that does not occur in the IMM estimation. Several mismatches like this one in the highly repetitive texture of the brick wall eventually may cause full tracking failure.

5.5.3 Camera Motion Model Identification

The IMM not only achieves better performance in processing the sequence, but also provides a tool to segment the sequence according to the dominant motion model. It is worth noting that this segmentation is based on sequence criteria as opposed to a classical pairwise motion model selection in geometrical vision.

In Figure 5.4 it can be seen that when there is a predominant model (stationary, rotating or general motion), the corresponding probability μ^i reaches a value close to 1, while the other model probabilities goes down close to zero — the IMM acts as a discrete selector here rather than a mixer. Only when there is a change between motion models are there periods with no clear dominant model and this is where the IMM proves its worth.

It has to be noted that models with lower acceleration noise are preferred unless the camera really undergoes a high acceleration motion. In fact the model with the highest acceleration has negligible probability indicating that it is not necessary for processing the current sequence. Although this unused model does require a computational overhead, its presence does not affect the accuracy of the solution nor jeopardize the matching by the size of the search regions for the predicted features — since its weight is always close to zero it is simply weighted out of all the calculations.

5.5.4 *Computational Cost Considerations*

Although the main advantage of the algorithm is its good tracking performance, clearly outperforming standard single model SLAM on complex sequences, it is also remarkable that the computational cost does not grow excessively. The cost of the IMM algorithm is essentially linear with the number of models since all filtering operations must be duplicated for each model. This is offset somewhat, as shown in section 5.5.2, by the fact that the search region ellipses are reduced in size in the IMM formulation and this makes the image processing work of feature matching cheaper.

5.6 Discussion

We have shown experimentally the advantages of the IMM filter when applied to EKF-based Structure from Motion. We are able to track sequences containing periods with no movement, and pure rotation and general motion at various dynamic levels, the system adapting automatically. In particular, while single model monocular SLAM is weak when bootstrapped with low parallax motions (still or rotating camera), the IMM formulation copes admirably by recognising the motion type.

The IMM formulation requires a computational overhead, but has extra benefits in producing smaller acceptance regions for the predicted measurements, improving outlier rejection, and being able to act as an automatic segmentation and labelling tool by identifying motion boundaries.

Chapter 6
Self-calibration

Abstract. Computer vision researchers have proved the feasibility of camera self-calibration –the estimation of a camera's internal parameters from an image sequence without any known scene structure. Nevertheless, all of the recent sequential approaches to 3D structure and motion estimation from image sequences which have arisen in robotics and aim at real-time operation (often classed as visual SLAM or visual odometry) have relied on pre-calibrated cameras and have not attempted online calibration. In this chapter, we present a sequential filtering algorithm for simultaneous estimation of 3D scene estimation, camera trajectory and full camera calibration from a sequence of fixed but unknown calibration. This calibration comprises the standard projective parameters of focal length and principal point along with two radial distortion coefficients.

6.1 Introduction

Camera self-calibration (or auto-calibration) is the process of estimating the internal parameters of a camera from a set of arbitrary images of a general scene. Self-calibration has several practical advantages over calibration with a special target. First, it avoids the onerous task of taking pictures of the calibration object; a task that may be difficult or even impossible if the camera is attached to a robot. Second, internal parameters of a camera may change either unintentionally (e.g. due to vibrations, thermical or mechanical shocks) or even intentionally in the case of a zooming camera. 3D estimation in this latter case could only be performed via self-calibration. Finally, inaccurate calibration (coming either from a poor calibration process or from changed calibration parameters) produces the undesirable effect of introducing bias in the estimation.

J. Civera et al.: Structure from Motion Using the Extended Kalman Filter, STAR 75, pp. 111–122.
springerlink.com

Although computer vision researchers have demonstrated the feasibility of self-calibration and despite all the advantages mentioned before, all of the recent sequential approaches to visual localisation and mapping –Visual Odometry and Visual SLAM– rely on a pre-calibrated camera. In this chapter, it is proposed a sequential SLAM-based algorithm that is able to sequentially estimate the structure of a scene, the trajectory of a camera and also its full calibration — including two coefficients of radial distortion. The only assumption made about the fixed camera calibration is that the skew is zero and the pixel aspect ratio is 1, a reasonable assumption in today's digital cameras.

Apart from all the advantages mentioned, self-calibration can also be considered essential for the development of practical systems. Vision systems in vacuum cleaners, autonomous vehicles or mobile phones cannot rely on end users to perform an accurate camera calibration for the system to work. Instead, it would be desirable that the visual estimation worked as soon as you install the software in your mobile phone or the engine of your car is started. For example, [Grasa *et al.* 2011], uses the real-time EKF SfM system described in this book to enhance visual perception in laparoscopic surgery. Calibrating the endoscope before every operation by medical staff would be impractical, being desirable that the system selfcalibrates in order to produce minimum interference and disturbances to the medical team.

The rest of the chapter is organised as follows: section 6.2 surveys prior work related to the approach presented here. Section 6.3 introduces the Sum of Gaussians (SOG) filter. In section 6.4 we detail our self-calibration algorithm using SOG. Section 6.5 presents real-image experiments that validate our approach. The conclusions and discussions about this chapter can be found in section 6.6.

6.2 Related Work

Traditionally, photogrammetric bundle adjustment has included camera calibration parameters — projective camera parameters and also distortion parameters — in order to refine a tight initial calibration guess and hence improve reconstruction accuracy.

Self-calibration allows the computation from scratch of projective calibration parameters: focal length, principal point, and skew; the computed calibration is readily usable or might be used as an initial guess for bundle adjustment refinement, and the refinement might include estimation of distortion parameters. The standard off-line self-calibration process is summarized as follows: first, matches along an uncalibrated sequence with possibly varying parameters are determined. Note that here, no assumptions about camera calibration — except that non-projective distortions are negligible — are applied. Then a projective reconstruction is computed; a potentially warped version of the ideal Euclidean reconstruction. If no more information about the camera taking the images is available then projective reconstruction is the best result that can be computed. However if some knowledge about calibration parameters is available — that they are constant, that there is zero skew, a known principal point or known aspect ratio — then this can be exploited to compute the rest of the unknown calibration parameters. Maybank and Faugeras demonstrated auto-calibration for the first time in 1992 [Maybank & Faugeras 1992, Faugeras *et al.* 1992]. Since then different methods for upgrading projective reconstruction to metric using partial knowledge about the camera calibration have been developed. A summary of all theses results is found in [Hartley & Zisserman 2004].

In spite of the vast amount of work related to autocalibration, approaches to these problem under a sequential Bayesian estimation framework are surprisingly few, and none of them performs a complete calibration. In [Azarbayejani & Pentland 1995] the authors propose for the first time the use of an EKF for sequential Bayesian estimation of unknown focal length. This is relevant seminal work but the 3D point parametrization is basic and this makes it difficult to deal with occlusion and feature addition and deletion. The approach of [Qian & Chellappa 2004] estimates a varying focal length assuming that the rest of the calibration parameters are known, and using a particle filter to deal with non-linearities.

In the context of visual SLAM, self-calibration of the internal parameters of a camera has not been previously discussed. [Solà *et al.* 2008] may be the closest related work, where a self-calibration algorithm for extrinsic camera parameters in large baseline stereo SLAM is proposed –relative locations of cameras in a stereo rig are easily decalibrated as baseline is increased. External self-calibration has also been tackled for multisensor systems, like [Scaramuzza *et al.* 2007] for a 3D laser and a camera. In this latest work, no artificial landmarks are needed but correspondences have to be manually identified. [Bryson *et al.* 2009] describes a multisensor system composed of GPS, IMU and monocular camera that autocalibrates IMU biases, IMU-camera relative angles and refines an initial guess of the intrinsic camera calibration. A general framework for self-calibration of non-visual sensor internal parameters –biases, gains, etc.– has also been addressed, e.g. in [Foxlin 2002].

Regarding the estimation techniques used in this work, the nonlinearity of the self-calibration problem has forced us to abandon the Extended Kalman Filter and adopt an approach more suitable for nonlinear systems: the Sum of Gaussians (SOG) filter [Alspach & Sorenson 1972]. This type of filter has already been used in SLAM [Piniés *et al.* 2006, Durrant-Whyte *et al.* 2003]. [Kwok *et al.* 2005] is of particular interest, as the combination of the Sum of Gaussians filter plus Sequential Probability Ratio Test they use to deal with the point initialization problem in monocular SLAM is the same it is used in this chapter for self-calibration purposes.

6.3 Sum of Gaussians (SOG) Filter

Within the SOG approach [Alspach & Sorenson 1972], the probability density function of the estimated parameters $p(\mathbf{x})$ is approximated by a weighted sum of multivariate Gaussians:

$$p(\mathbf{x}) = \sum_{i=1}^{n_g} \alpha^{(i)} \mathscr{N}\left(\hat{\mathbf{x}}^{(i)}, \mathbf{P}^{(i)}\right) , \tag{6.1}$$

where n_g stands for the number of Gaussians, $\hat{\mathbf{x}}^{(i)}$ and $\mathbf{P}^{(i)}$ are the mean and covariance matrix for each Gaussian and $\alpha^{(i)}$ represents the weighting factors, which should obey $\sum_{i=1}^{n_g} \alpha^{(i)} = 1$ and $\alpha^{(i)} \geq 0$.

This Sum of Gaussians probability density function evolves as follows: at every step, when new measurements arrive, each one of the Gaussians is updated with the standard prediction-update Extended Kalman Filter equations. The central part of the SOG algorithm is, then, a bank of EKF filters running in parallel. This bank of EKF filters is illustrated in Figure 6.1.

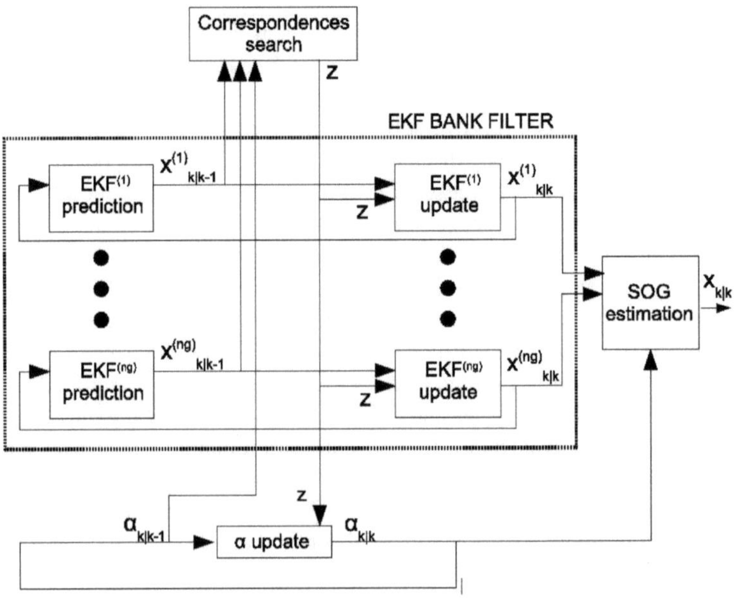

Fig. 6.1. Scheme of the Sum of Gaussians (SOG) filter.

Weighting factors $\alpha^{(i)}$ are also updated at every step k using this formula:

$$\alpha_k^{(i)} = \frac{\alpha_{k-1}^{(i)} \mathscr{N}\left(v_k^{(i)}, \mathbf{S}_k^{(i)}\right)}{\sum_{j=1}^{n_g} \alpha_{k-1}^{(j)} \mathscr{N}\left(v_k^{(j)}, \mathbf{S}_k^{(j)}\right)} ; \tag{6.2}$$

where $v_k^{(i)}$ and $\mathbf{S}_k^{(i)}$ are the EKF innovation vector and its covariance matrix respectively. The innovation vector for each EKF is computed as the difference between the actual measurements $\mathbf{z}_k$ and the predicted measurements $\mathbf{h}_k^{(i)}$. The predicted measurements $\mathbf{h}_k^{(i)}$ result from applying the measurement model equations to the state vector $\hat{\mathbf{x}}_k^{(i)}$

$$\nu_k^{(i)} = \mathbf{z}_k - \mathbf{h}_k^{(i)}, \quad \mathbf{h}_k^{(i)} = \mathbf{h}\left(\hat{\mathbf{x}}_k^{(i)}\right). \tag{6.3}$$

The innovation covariance is obtained propagating each EKF covariance $\mathbf{P}_k^{(i)}$ through the measurement equations and adding the covariance of the zero-mean image noise $\mathbf{R}_k^{(i)}$

$$\mathbf{S}_k^{(i)} = \mathbf{H}_k^{(i)} \mathbf{P}_k^{(i)} \mathbf{H}_k^{(i)\top} + \mathbf{R}_k^{(i)}, \quad \mathbf{H}_k^{(i)} = \left.\frac{\partial \mathbf{h}}{\partial \mathbf{x}}\right|_{\hat{\mathbf{x}}_k^{(i)}}. \tag{6.4}$$

Finally, an overall mean and covariance for the whole filter can be computed as follows:

$$\begin{aligned} \hat{\mathbf{x}}_k &= \sum_{i=1}^{n_g} \alpha_k^{(i)} \hat{\mathbf{x}}_k^{(i)} \\ \mathbf{P}_k &= \sum_{i=1}^{n_g} \alpha_k^{(i)} \left(\mathbf{P}_k^{(i)} + \left(\hat{\mathbf{x}}_k^{(i)} - \hat{\mathbf{x}}_k\right)\left(\hat{\mathbf{x}}_k^{(i)} - \hat{\mathbf{x}}_k\right)^\top\right). \end{aligned} \tag{6.5}$$

These latest overal mean and covariance are used for visualization purposes in our experiments. Nevertheless, notice (graphically in Figure 6.1) that this is the only purpose of the overall mean and covariance as they are not involved either in the filter bank or in the evolution of the weighting factors.

From this brief introduction, the two fundamental advantages of the SOG filter over the EKF can be intuitively introduced. First, notice that any probability density function can be reasonably approximated by a weighted Sum of Gaussians if we make the number of Gaussians n_g high enough. So, the usual EKF assumption of Gaussian PDF does not need to hold for the SOG filter. Second, and more importantly for this work, as we increase the number of Gaussians the uncertainty $\mathbf{P}^{(i)}$ for each Gaussian becomes smaller, favoring linearity.

6.4 Self-calibration Using SOG Filtering

6.4.1 State Vector Definition

In order to estimate 3D scene structure and camera location and calibration the SOG state vector $\mathbf{x}$, –and hence every EKF state vector $\mathbf{x}^{(i)}$ that composes the filter bank– will contain a set of camera parameters $\mathbf{x}_C$ and a set of parameters $\mathbf{x}_M$ representing each estimated point $\mathbf{y}_j$.

$$\mathbf{x} = \left(\mathbf{x}_C^\top, \mathbf{x}_M^\top\right)^\top, \quad \mathbf{x}_M = \left(\mathbf{y}_1^\top, \ldots, \mathbf{y}_n^\top\right)^\top \tag{6.6}$$

Mapped points $\mathbf{y}_j$ are first coded in inverse depth coordinates and converted into Cartesian (XYZ) coordinates if and when their measurement equation becomes linear, as detailed in chapter 3.

$$\mathbf{y}_\rho = (X_c, Y_c, Z_c, \theta, \phi, \rho)^\top, \quad \mathbf{y}_{\text{XYZ}} = (X, Y, Z)^\top \tag{6.7}$$

The camera part of the state vector $\mathbf{x}_C$, as the key difference from previous work, now includes the internal calibration parameters to estimate: the focal length f, the principal point coordinates C_x and C_y and the parameters modelling radial distortion κ_1 and κ_2.

$$\mathbf{x}_C = \left(\mathbf{x}_K^\top, \mathbf{x}_v^\top\right)^\top; \quad \mathbf{x}_K = (f, C_x, C_y, \kappa_1, \kappa_2)^\top,$$

$$\mathbf{x}_v = \begin{pmatrix} \mathbf{r}^{WC} \\ \mathbf{q}^{WC} \\ \mathbf{v}^{W} \\ \omega^{C} \end{pmatrix}. \tag{6.8}$$

The camera motion model is described in section 3.2.1; and the projection model is the one in section 3.3. Nevertheless, it is important to remark that calibration parameters in this model that were assumed to be known up to this chapter are now taken from the state vector.

6.4.2 *Pruning of Gaussians with Low Weight*

As it can be assumed that the final estimation result will be unimodal, Gaussians that repeteadly obtain low weighting factors can be pruned reducing the computational cost. To do this, it is adopted the proposal in [Kwok *et al.* 2005], which makes use of the Sequential Probability Ratio Test (SPRT) [Wald 1945]. Experiments have shown that SPRT achieves a high reduction rate while maintaining similar performance.

For each Gaussian i in the SOG filter, the null hypothesis H_0 is that such Gaussian correctly represents the true state and the alternative hypothesis H_1 that the Gaussian does not represent the true state. At every step k, the null hypothesis is accepted if

$$\prod_{t=1}^{k} \frac{\mathcal{L}_t^{(i)}(H_0)}{\mathcal{L}_t^{(i)}(H_1)} > A, \tag{6.9}$$

and the alternative hypothesis is accepted (meaning that Gaussian i can be pruned) if

$$\prod_{t=1}^{k} \frac{\mathcal{L}_t^{(i)}(H_0)}{\mathcal{L}_t^{(i)}(H_1)} < B, \tag{6.10}$$

where $\mathscr{L}_t^{(i)}(H_0)$ and $\mathscr{L}_t^{(i)}(H_1)$ are the likelihoods of the data under hypothesis H_0 and H_1 at frame t. These likelihoods are computed as follows:

$$\mathscr{L}_t^{(i)}(H_0) = \mathscr{N}\left(\mathbf{v}_t^{(i)}, \mathbf{S}_t^{(i)}\right) \tag{6.11}$$

$$\mathscr{L}_t^{(i)}(H_1) = \sum_{j=1;j\neq i}^{n_g} \alpha^{(j)\prime} \mathscr{N}\left(\mathbf{v}_t^{(j)}, \mathbf{S}_t^{(j)}\right) \tag{6.12}$$

$$\alpha^{(j)\prime} = \frac{\alpha^{(j)}}{\sum_{k=1;k\neq i}^{n_g} \alpha^{(k)}} \, . \tag{6.13}$$

Thresholds A and B are approximated by the so-called *Wald Boundaries* [Wald 1945] $A = \frac{1-\alpha_b}{\alpha_a}$ and $B = \frac{\alpha_b}{1-\alpha_a}$, where α_a and α_b are the probabilities of type I and type II errors.

6.5 Experimental Results

Two experiments have been carried to test the performance of the algorithm. The design of the SOG filter, which is the same for both experiments, it is explained here previous to the experimental results.

An interval for the focal length f from around 100 pixels to around 600 pixels is considered to be the usual range for cameras used in robotics. It has been experimentally found that the projection measurement equation from section 3.3 is fairly linear in intervals of 30 pixels. So, in order to estimate the focal length, the full range of possible focal length values is divided into 18 Gaussians with standard deviations of 7.5 pixels and separation between means of 30 pixels. Figure 6.2 shows the resulting probability distribution function.

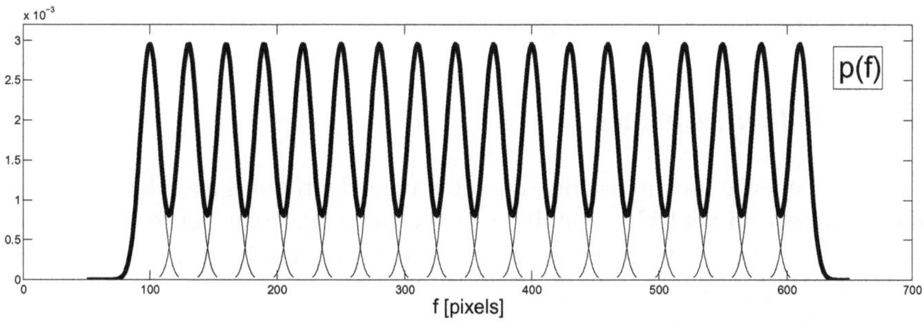

Fig. 6.2. Probability density function considered for the focal length.

A similar procedure applies for κ_1 and κ_2. It is considered that usual values for these parameters go from 0 (no radial distortion) to 0.08mm^{-2} and 0.018mm^{-4}

respectively. The projection model is approximately linear if these two variation ranges are divided into 2 intervals for κ_1 and 3 for κ_2. The resulting probability density functions for radial distortion parameters can be seen in Figure 6.3.

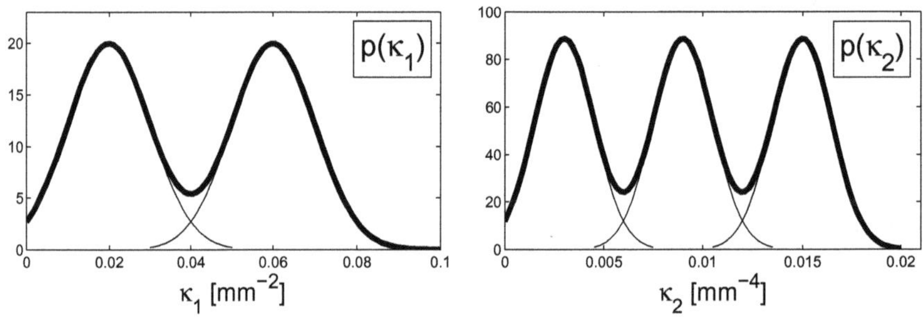

Fig. 6.3. Probability density function for distortion parameters κ_1 and κ_2.

The final SOG filter will be composed of all possible combinations of the above divisions, that is $18 \times 2 \times 3 = 108$ filters. It is worth mentioning here that the naive approach of considering all combinations of parameters is used here only to demonstrate the performance of the algorithm in the most general case. In a practical setting, it is known that a camera with large focal length will have negligible radial distortion and that small focal lengths corresponds to high radial distortion lenses. Hence, the number of the filters could be greatly reduced by observing the most usual combinations of calibration parameters.

Finally, regarding the optical centre coordinates C_x and C_y; as the measurement equation is linear for those parameters, they are coded with one single Gaussian. The optical centre is assumed to be a maximum of 10 pixels from the centre of the image. For a 320×240 image, this results in a bidimensional Gaussian whose mean is $[160, 120]$ and whose standard deviations are 3.3 pixels in each coordinate.

6.5.1 Indoor Sequence

The first sequence used to test the self-calibration algorithm is an indoor sequence taken with a hand-held 320×240 IEEE1394 camera in a computer room. The purpose of this experiment is to test the accuracy of the proposed algorithm, comparing its results with an offline calibration.

Figure 6.4 shows three frames of the sequence, one at the beginning, the second in the middle and the last frame of the sequence, and with the 3D estimation at each instant. The evolution of the calibration parameters estimation over the sequence can be observed in Figure 6.5. The same figure also shows the number of Gaussians in the SOG filter at each step. Notice the steep decrease in the first steps of the estimation, and how after image 120 of the sequence the SOG filter is composed of only one filter, becoming an EKF. Table 6.1 details the initial and final values of the estimation with a 99% confidence interval and the offline calibration values for a better visualization of the accuracy of our self-calibration results. Notice that although initial values cover a wide range of variation for the parameters, the SOG ends up with a tight and consistent estimation for all of them.

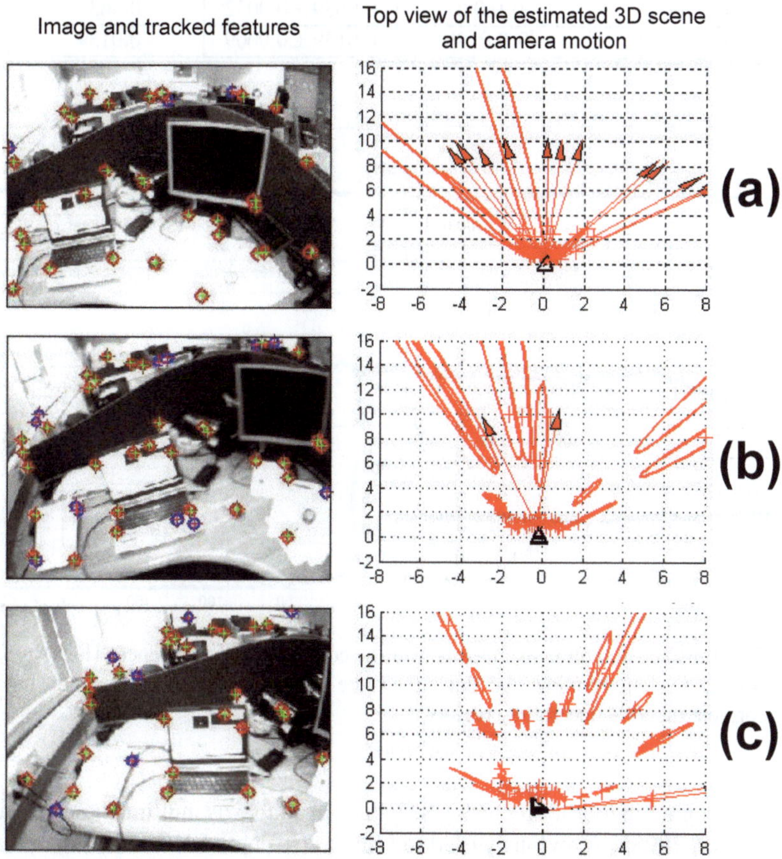

Fig. 6.4. Images and top-down view 3D estimation for frames #20 (a), #80 (b) #260 (c), which is the last frame of the sequence.

6.5.2 *Loop-Closing Sequence*

Loop-Closing is a standard benchmark in SLAM to test the validity of an estimation algorithm: when a sensor revisits known areas, the estimation error should be small enough for the algorithm to recognize previous mapped landmarks.

Table 6.1. Calibration results for indoor sequence

	Initial SOG Interval	Final SOG Estimation	Offline Calibration
$f[pixels]$	$[100,610]$	193.0 ± 1.9	194.1
$C_x[pixels]$	$[150,170]$	161.6 ± 2.3	160.2
$C_y[pixels]$	$[110,130]$	127.0 ± 2.4	128.9
$\kappa_1[mm^{-2}]$	$[0,0.08]$	0.0639 ± 0.0032	0.0633
$\kappa_2[mm^{-4}]$	$[0,0.018]$	0.0139 ± 0.0009	0.0139

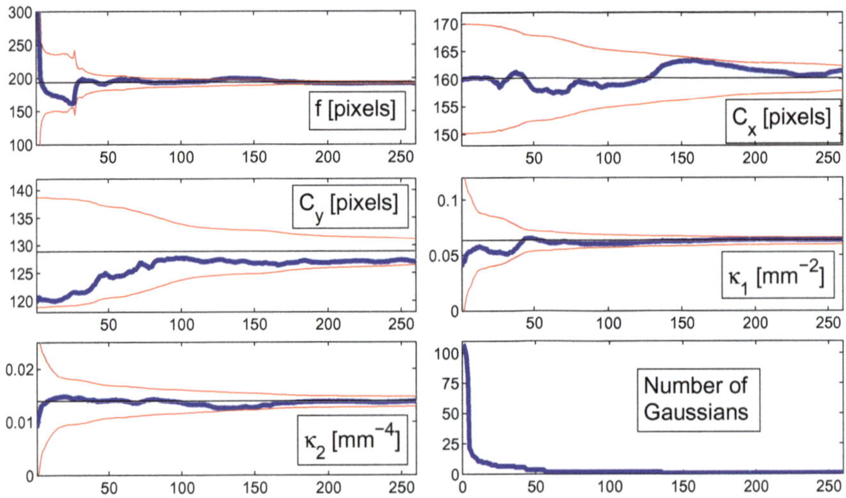

Fig. 6.5. Estimated calibration parameters over the computer room sequence. Thick blue line is the estimated value, the horizontal black line is the offline calibration value and the red thin lines represent the 99% uncertainty region.

A challenging indoor loop-closing sequence available as multimedia material in [Civera *et al.* 2008a] –previously used to test inverse depth EKF monocular SLAM with a calibrated camera in section 3.8.3– has been used in this experiment. The estimated calibration values are accurate enough to close the loop. Figure 6.6 shows three representative frames of the sequence and their estimated scene, including the loop closing frame.

As we show in Table 6.2 and in Figure 6.7, the estimated calibration is close to the offline calibration, but in a slightly over-confident manner. When compared with the previous one, this experiment presents more difficult linearization issues because uncertainty increases when the camera explores new areas, and increases in uncertainty imply more non-linear effects. Besides, the fixed model for the calibration parameters implies a monotonic uncertainty reduction that becomes unrealistic after processing several hundred of images.

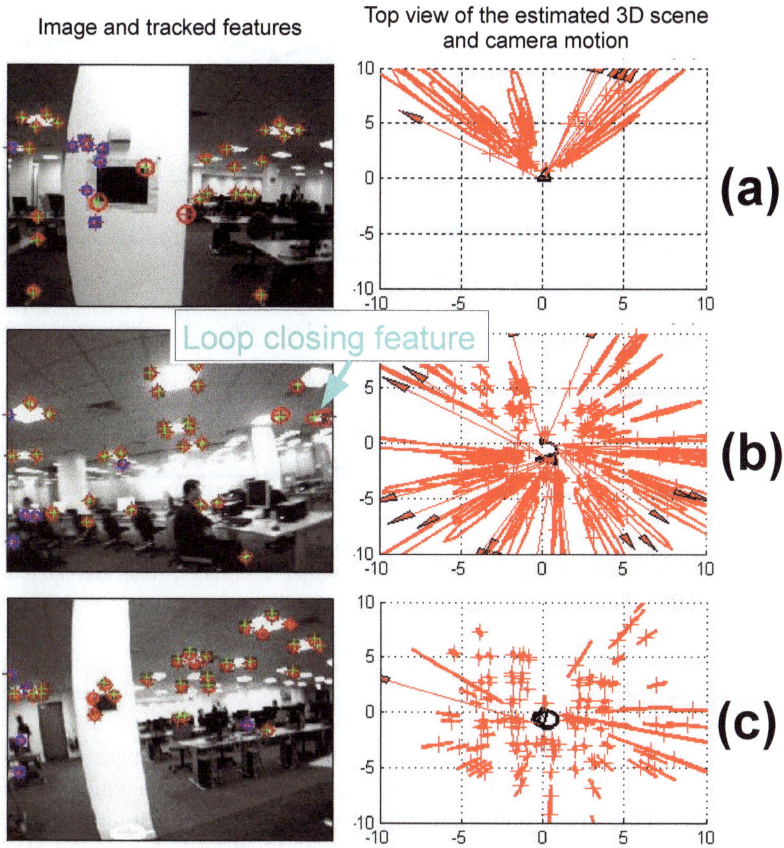

Fig. 6.6. (a) Image and 3D estimation at frame 60. (b) Image and 3D estimation at frame 330 of the sequence, when first loop-closure feature (signaled in the image) is detected. (c) Image and 3D estimation at frame 670, the last one of the sequence.

Table 6.2. Calibration results for the loop closing sequence.

	Initial SOG Interval	Final SOG Estimation	Offline Calibration
$f[pixels]$	$[100,610]$	195.0 ± 0.4	196.9
$C_x[pixels]$	$[150,170]$	159.6 ± 1.0	153.5
$C_y[pixels]$	$[110,130]$	133.9 ± 1.0	130.8
$\kappa_1[mm^{-2}]$	$[0,0.08]$	0.0652 ± 0.0019	0.0693
$\kappa_2[mm^{-4}]$	$[0,0.018]$	0.0132 ± 0.0005	0.0109

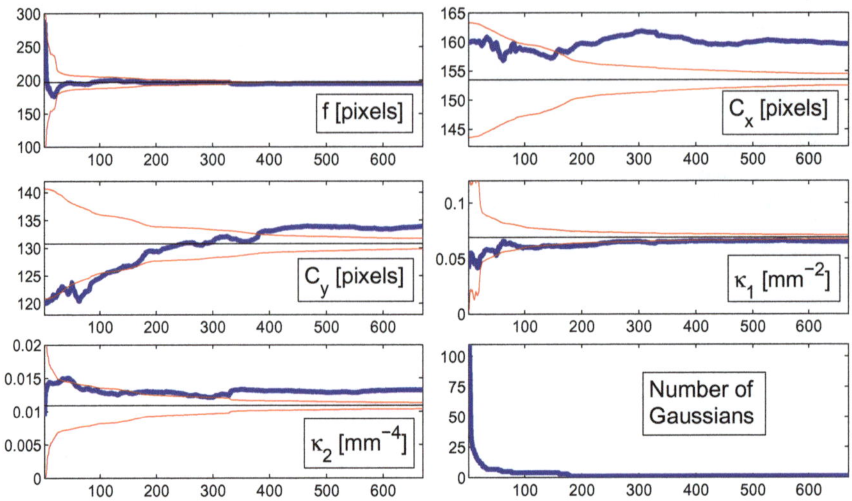

Fig. 6.7. Estimated calibration parameters over the loop closing sequence. Thick blue line is the estimated value, the horizontal black line is the offline calibration value and the red thin lines represent the 99% uncertainty region.

6.6 Discussion

This chapter presents an algorithm that fully auto-calibrates a camera within a filtering framework, the only input being a sequence of images from a moving uncalibrated camera. Due to non-linearities introduced by the estimation of calibration parameters, a Sum of Gaussian filter is used to divide the whole non-linear range of variation into small almost-linear pieces. The SOG approach uses several filters in the first steps of the estimation to cover all of these almost-linear hypothesis. A pruning algorithm has been added that cuts Gaussians whose weighting factors are low and reduces the SOG filter to a simple EKF in a few steps so complexity is reduced after an initial computation overhead. As the multiple Gaussians have to be kept only at initial stages when the map size is small, we expect the computational complexity to be low enough to achieve real time performance.

Experimental results with real-images show that an accurate and consistent camera calibration is achieved for a waggling motion in an indoor sequence. A loop closure has been successfully performed, achieving calibration values close to offline calibration, what is a remarkable achievement, though the estimation is somewhat inconsistent due to non-linearities and to the unrealistic monotonic uncertainty reduction that EKF produces when dealing with static parameters.

Regarding future lines of work, an interesting one would be to analyze how this self-calibration algorithm behaves with respect to degenerate camera motion. Also, being already demonstrated that sequential camera self-calibration is feasible for a camera with fixed unknown parameters, next natural step is to deal with varying calibration parameters.

Chapter 7
Conclusions

Abstract. This chapter presents the main conclusions and summarizes the content of the book. The algorithms, models and methods presented in the previous chapters cover the main topics in sequential SfM or monocular SLAM: a projective point model, an efficient and robust search for correspondences and altorithms for model selection and internal self-calibration. Together, the contributions presented in the different chapters of the book form a robust system for sequential scene and camera motion estimation potentially able to deal with *any* image sequence in real-time at 30 frames per second.

The sequential 3D camera motion and scene structure from a monocular video stream has become a very relevant topic in diverse areas like robotics and augmented reality (AR). In the robotic case, the estimation of a model of the environment and the position of the robot into it allows to interact and manipulate the scene without collision. From the wide array of available sensing modalities, the monocular camera seems to hold a promising future in robotics due to its compacity, low cost and accuracy and richness of information. The extraction of the geometric information contained in a video sequence in an efficient and sequential manner seems to be the current challenge in both areas. Apart from the already mentioned robotic and AR fields, the growing importance of video data in the Internet [Yang & Toderici 2011] and the popularization of the smartphones could open new and exciting application scenarios. Also, and now that the real-time 3D scene and motion estimation from natural videos has become commonplace, e.g. [Klein & Murray 2009, Sibley *et al.* 2009, Strasdat *et al.* 2011], the authors see as a very promising line the sequential and real-time joint estimation of geometry and high-level information (e.g., geometry and object recognition in [Civera *et al.* 2011, Bao & Savarese 2011])

The authors believe that the pure monocular case is an excellent test bed for more general visual sensing systems in this research scenario. It is clear that monocular SLAM algorithms improve their performance if they are adapted to a more evolved sensing modality –like stereo or visual plus inertial sensing. Nevertheless, the data

J. Civera et al.: Structure from Motion Using the Extended Kalman Filter, STAR 75, pp. 123–125.
springerlink.com

provided by these latest systems very frequently hides the real challenges of visual sensing; challenges that are easily spotted in the pure monocular case. Once a solution is developed in this minimalistic sensing scenario, the transfer to a more sophisticated system will improve its performance. For example, the inverse depth coding for feature initialization presented in chapter 3 and developed for the monocular case has been successfully transferred to the stereo case [Paz *et al.* 2008a] and for inertial plus monocular sensing [Piniés *et al.* 2007]. Also, the monocular input is the only choice in some cases like Internet videos or movies; and it is always the cheapest one.

The contents of the present book have presented a consistent formulation for real-time Structure from Motion from uncalibrated images using the Extended Kalman Filter (EKF). The EKF has been one of the fundamental estimation techniques in robotic SLAM since [Smith & Cheeseman 1986b] due to its sequential and probabilistic processing of the sequence; issues that were considered essential for a robotic application. Sensorial data from robots comes in the form of a stream and its information needs to be inserted in the control loop, hence the sequential processing. As a robot is intended to interact with a physical environment, the uncertainty of the estimation should be taken into account in order to avoid collisions. Those early robotic papers mostly used laser sensors. Early work on the filtering of visual sequences can be found in [Azarbayejani & Pentland 1995]. Nevertheless, the computer vision community was more focused on the formalization of the geometry of separated views than the processing of video sequences. The so-called Structure from Motion problem occupied the researchers for several decades, and it is summarized in the excellent [Hartley & Zisserman 2004].

The chapters of this book unify both the perspectives of the robotics and computer vision community on the 3D estimation from sequences. First, the tipycally robotic estimation of the uncertainty holds a central role in the sequence processing, allowing a boost in the efficiency and performance of the algorithms. On the other hand, the most relevant topics in the Structure from Motion literature are addressed in this book under a filtering perspective.

More specifically, these relevant topics are the following: First, the need for a projective camera model is borrowed from the computer vision community and transferred to the SLAM problem. This allows a straightforward representation of low-parallax points, both distant and newly initialized ones –solving the so-called *initialization* problem. Apart from being adequate to represent low-parallax points, the inverse depth parametrization proposed in chapter 3 holds a high degree of linearity in its projection equation which leads to a nice EKF performance. A mosaicing application, typical from the computer vision community, is shown to benefit from a SLAM-type processing and presents for the first time real-time drift-free results in chapter 2. This line of research is continuated in [Lovegrove & Davison 2010].

Chapter 4 presents an efficient spurious rejection algorithm, where the typically used in computer vision *RANSAC* benefits from the uncertainty propagation of the EKF showing large computational savings without performance degradation. The combination of the projective inverse depth parametrization and an efficient spurious rejection result in a highly accurate estimation for large experiments where the camera translates hundreds of metres.

The two latest chapters deal with degenerate motions and self-calibration; two topics largely discussed in the computer vision community and that are introduced here to the robotic community. Again, the uncertainty of the estimation keeps a key role in both approaches. Degenerate motions are dealt by selecting the motion model based on probabilities without the need for terms penalizing complex models. Degenerate motions, e.g., pure rotation or a small translation compared with the scene depth, are not uncommon in a robotic application and should be explicitly addressed to avoid inconsistent estimations.

The self-calibration is considered by the authors as an advisable feature in most SLAM systems in order to avoid an onerous pre-calibration with a target. Such pre-calibration could be inaccurate when performed by an inexperienced user and even be very inconvenient in certain configurations. A clear example can be the system in [Grasa *et al.* 2011], that aims to perform SLAM from endoscopic images to help surgeons. Such system should produce a minimum interference with the work of the medical staff and the system should work with the highest degree of automony, being self-calibration a must.

The application of the concepts in this book expands then the array of sequences that may be processed by the monocular SLAM algorithms, increases the accuracy and robustness of the estimation and has proven to run in real-time at 30 frames per second in a standard laptop for small local maps. From a computer vision perspective, the research presented here follows the path initiated 30 years ago aiming to an online motion and structure estimation from an image sequence and contributes by linking the most relevant topics in wide baseline Structure from Motion with the online estimation from short-interframe-motion image sequences. From the point of view of the robotic research the approaches presented in this book form a practical system that could be said to perform *out-of-the-box monocular SLAM*, that is, SLAM from a single camera with very few constraints of use. The discussions in the book also have the value of providing a deeper understanding of the camera as a sensor in the field of robotics. Finally, and although further research in such an active area as 3D vision is expected, the presented system already fits the accuracy, robustness and real-time requirements for a number of potential applications and hence can be taken as a basic sensing piece for complex application-driven systems.

Appendix A
Implementation Details

A.1 Extended Kalman Filter

The Kalman Filter –and its version for non-linear systems, the Extended Kalman Filter– are probably the best studied implementation of Bayesian filtering and the first filtering solution to be succesfully applied to the online SLAM estimation problem [Smith *et al.* 1987].

The Kalman filter recursively estimates a probability distribution function over the unknown parameters of a state vector $\mathbf{x}$ from measurements gathered by a sensor and the dynamical model of such state. The estimation process follows two steps. In the first one, the probability distribution function $p\left(\mathbf{x}_{k-1}\right)$ from step $k-1$ is updated to step k based on the probabilistic dynamic model of the system $p\left(\mathbf{x}_{k|k-1}|\mathbf{x}_{k-1|k-1},\mathbf{u}_k\right)$ in the so-called *prediction* step

$$p\left(\mathbf{x}_{k|k-1}\right)=\int p\left(\mathbf{x}_{k|k-1}|\mathbf{x}_{k-1|k-1},\mathbf{u}_k\right)p\left(\mathbf{x}_{k-1}\right)d\mathbf{x}_{k-1}\;. \tag{A.1}$$

After measurements $\mathbf{z}_k$ are collected, they are fused with the probability distribution function from the prediction step using Bayes' rule in the *update* step

$$p\left(\mathbf{x}_{k|k}\right)=\eta\, p\left(\mathbf{z}_k|\mathbf{x}_{k|k-1}\right)p\left(\mathbf{x}_{k|k-1}\right)\;, \tag{A.2}$$

where $p\left(\mathbf{z}_k|\mathbf{x}_{k|k-1}\right)$ stands for the probabilistic measurement model of the system, that has be known in advance; and η is the normalization constant that converts $p\left(\mathbf{x}_{k|k}\right)$ into a probability distribution function. The algorithm require the state probability distribution at the initial step $p\left(\mathbf{x}_0\right)$ to be known.

The Kalman Filter, initially proposed in [Kalman 1960], makes three assumptions: linear dynamic and measurement model with Gaussianly distributed noise; and Gaussian initial probability distribution at time $k=0$. Under these assumptions, the a posteriori distribution over the estimated parameters is also a multivariate Gaussian $\mathbf{x}\sim\mathscr{N}\left(\hat{\mathbf{x}},\mathbf{P}\right)$.

Most of the systems in the real world show a certain degree of non-linearity. The Extended Kalman Filter (EKF) relaxes the linearity assumption by linearizing the dynamic and measurement model in the mean value at every step of the estimation. The Extended Kalman Filter does not give us then the real a posteriori probability distribution function, but only a Gaussian approximation of it. The higher the degree of linearity in the dynamic and measurement models and the more Gaussian the noise, the better the EKF will perform.

The prediction step using the dynamic model of the system, described in its more general form in equation A.1, makes the mean $\hat{\mathbf{x}}$ and covariance $\mathbf{P}$ of the Gaussian probability in the EKF evolve in the following manner

$$\hat{\mathbf{x}}_{k|k-1} = \mathbf{f}\left(\hat{\mathbf{x}}_{k|k-1}, \mathbf{u}_k\right) \ , \tag{A.3}$$

$$\mathbf{P}_{k|k-1} = \mathbf{F}_k \mathbf{P}_{k|k-1} \mathbf{F}_k^\top + \mathbf{G}_k \mathbf{Q}_k \mathbf{G}_k^\top \ ; \tag{A.4}$$

where $\mathbf{f}\left(\hat{\mathbf{x}}_{k|k-1}, \hat{\mathbf{u}}_k\right)$ is the nonlinear equation modeling the dynamic evolution of the system, $\mathbf{u}_k$ the input given to the system, $\mathbf{F}_k$ the derivatives of the dynamic model by the state vector, $\mathbf{Q}_k$ the input covariance and $\mathbf{G}_k$ the derivatives of the dynamic model by the input parameters. $\mathbf{I}$ is an identity matrix of an appropriate size.

The update step via Bayes' rule in equation A.2 is, in the EKF case

$$\hat{\mathbf{x}}_{k|k} = \hat{\mathbf{x}}_{k|k-1} + \mathbf{K}_{k|k-1}\left(\mathbf{z}_k - \mathbf{h}\left(\hat{\mathbf{x}}_{k|k-1}\right)\right) \tag{A.5}$$

$$\mathbf{P}_{k|k} = \left(\mathbf{I} - \mathbf{K}\mathbf{H}_{k|k-1}\right)\mathbf{P}_{k|k-1} \tag{A.6}$$

$$\mathbf{K}_{k|k-1} = \mathbf{P}_{k|k-1}\mathbf{H}_{k|k-1}^\top\left(\mathbf{H}_{k|k-1}\mathbf{P}_{k|k-1}\mathbf{H}_{k|k-1}^\top + \mathbf{R}_k\right)^{-1} \ ; \tag{A.7}$$

where $\mathbf{z}_k$ are measurements gathered at step k, $\mathbf{h}\left(\hat{\mathbf{x}}_{k|k-1}\right)$ the function that defines the sensor measurement model, $\mathbf{H}_{k|k-1}$ the derivatives of the measurement function by the state vector and $\mathbf{R}_k$ the covariance of the measurement noise. The matrix $\mathbf{K}_{k|k-1}$ is called the filter gain and weights the information from the prior knowledge and the measurements according to their respective covariances.

For the practical implementation of an EKF filter, it is then required to know: first, wich parameters are included in the state vector and its probabilistic dynamic model. Second, which sensor we are using, what are our measurements and the probabilistic measurement model. Based on that, derivatives of both models can be computed and the EKF steps described in this section can be coded. The next sections detail the models and derivatives for the particular implementation of the monocular EKF SLAM system used along this book.

A.2 Calibrated EKF-Based SfM

A.2.1 Dynamic Model and Derivatives

The camera motion is considered smooth and modeled with a constant velocity model with a zero-mean Gaussian acceleration noise $\mathbf{n} = \begin{pmatrix} \mathbf{a}^W \\ \alpha^C \end{pmatrix}^\top$. As described in previous chapters, the camera state vector is composed of its position $\mathbf{r}^{WC}$ with respect to a world reference frame W, the quaternion representing its orientation $\mathbf{q}^{WC}$ and the linear and angular velocities $\mathbf{v}^W$ and ω^C –this latest one in the camera frame C.

$$\mathbf{x}_C = \begin{pmatrix} \mathbf{r}^{WC} \\ \mathbf{q}^{WC} \\ \mathbf{v}^W \\ \omega^C \end{pmatrix} \tag{A.8}$$

In this model, the camera states are modified at every step by impulses of linear velocity $\mathbf{V}^W = \mathbf{a}^W \Delta t$ and angular velocity $\Omega^C = \alpha^C \Delta t$ produced by a Gaussian acceleration noise:

$$\mathbf{x}_{C_{k+1}} = \begin{pmatrix} \mathbf{r}^{WC}_{k+1} \\ \mathbf{q}^{WC}_{k+1} \\ \mathbf{v}^W_{k+1} \\ \omega^C_{k+1} \end{pmatrix} = \mathbf{f}_v\left(\mathbf{x}_{C_k}, \mathbf{n}\right) = \begin{pmatrix} \mathbf{r}^{WC}_k + \left(\mathbf{v}^W_k + \mathbf{V}^W_k\right)\Delta t \\ \mathbf{q}^{WC}_k \times \mathbf{q}\left(\left(\omega^C_k + \Omega^C\right)\Delta t\right) \\ \mathbf{v}^W_k + \mathbf{V}^W \\ \omega^C_k + \Omega^C \end{pmatrix}. \tag{A.9}$$

$\mathbf{q}\left(\left(\omega^C_k + \Omega^C\right)\Delta t\right)$ stands for the quaternion corresponding to the rotation given by $\left(\omega^C_k + \Omega^C\right)\Delta t$; and $\times$ for the quaternion product.

The derivatives of this dynamic model by the state $\mathbf{F} = \frac{\partial \mathbf{f}_v}{\partial \mathbf{x}_C}$ and by the Gaussian noise $\mathbf{G} = \frac{\partial \mathbf{f}_v}{\partial \mathbf{n}}$ are computed as follows:

$$\frac{\partial \mathbf{f}_v}{\partial \mathbf{x}_C} = \begin{pmatrix} \mathbf{I} & \mathbf{0} & \Delta t \mathbf{I} & \mathbf{0} \\ \mathbf{0} & \frac{\partial \mathbf{q}^{WC}_{k+1}}{\partial \mathbf{q}^{WC}_k} & \mathbf{0} & \frac{\partial \mathbf{q}^{WC}_{k+1}}{\partial \omega^C_k} \\ \mathbf{0} & \mathbf{0} & \mathbf{I} & \mathbf{0} \\ \mathbf{0} & \mathbf{0} & \mathbf{0} & \mathbf{I} \end{pmatrix} \tag{A.10}$$

$$\frac{\partial \mathbf{f}_v}{\partial \mathbf{n}} = \begin{pmatrix} \Delta t \mathbf{I} & \mathbf{0} \\ \mathbf{0} & \frac{\partial \mathbf{q}^{WC}_{k+1}}{\partial \Omega^C} \\ \mathbf{I} & \mathbf{0} \\ \mathbf{0} & \mathbf{I} \end{pmatrix} \tag{A.11}$$

Let the quaternions in the previous formula be renamed as $\mathbf{q}^3 = \mathbf{q}_{k+1}^{WC}$, $\mathbf{q}^2 = \mathbf{q}_k^{WC}$ and $\mathbf{q}^1 = \mathbf{q}\left(\left(\omega_k^C + \Omega^C\right)\Delta t\right)$ and the components of each quaternion represented as $\mathbf{q} = (q_0, q_1, q_2, q_3)^\top$. The partial derivative $\frac{\partial \mathbf{q}_{k+1}^{WC}}{\partial \mathbf{q}_k^{WC}}$ is computed from the quaternion product formula

$$\frac{\partial \mathbf{q}_{k+1}^{WC}}{\partial \mathbf{q}_k^{WC}} = \frac{\partial \mathbf{q}^3}{\partial \mathbf{q}^2} = \begin{pmatrix} q_0^1 & -q_1^1 & -q_2^1 & -q_3^1 \\ q_1^1 & q_0^1 & q_3^1 & -q_2^1 \\ q_2^1 & -q_3^1 & q_0^1 & q_1^1 \\ q_3^1 & q_2^1 & -q_1^1 & q_0^1 \end{pmatrix} . \tag{A.12}$$

And the partial derivative $\frac{\partial \mathbf{q}_{k+1}^{WC}}{\partial \omega_k^C}$ can be computed via the chain rule as

$$\frac{\partial \mathbf{q}_{k+1}^{WC}}{\partial \omega_k^C} = \frac{\partial \mathbf{q}_{k+1}^{WC}}{\partial \mathbf{q}\left(\left(\omega_k^C + \Omega^C\right)\Delta t\right)} \frac{\partial \mathbf{q}\left(\left(\omega_k^C + \Omega^C\right)\Delta t\right)}{\partial \omega_k^C} ; \tag{A.13}$$

where $\frac{\partial \mathbf{q}_{k+1}^{WC}}{\partial \mathbf{q}\left(\left(\omega_k^C + \Omega^C\right)\Delta t\right)}$ comes again from the quaternion product formula

$$\frac{\partial \mathbf{q}_{k+1}^{WC}}{\partial \mathbf{q}\left(\left(\omega_k^C + \Omega^C\right)\Delta t\right)} = \frac{\partial \mathbf{q}^3}{\partial \mathbf{q}^1} = \begin{pmatrix} q_0^2 & -q_1^2 & -q_2^2 & -q_3^2 \\ q_1^2 & q_0^2 & -q_3^2 & q_2^2 \\ q_2^2 & q_3^2 & q_0^2 & -q_1^2 \\ q_3^2 & -q_2^2 & q_1^2 & q_0^2 \end{pmatrix} \tag{A.14}$$

and $\frac{\partial \mathbf{q}\left(\left(\omega_k^C + \Omega^C\right)\Delta t\right)}{\partial \omega_k^C}$ can be computed from the conversion formula from a rotation vector to a quaternion representation. This conversion formula is

$$\mathbf{q}(\omega) = \left(\cos\frac{\theta}{2}, \frac{\omega^\top}{\theta}\sin\frac{\theta}{2}\right)^\top \tag{A.15}$$

$$\theta = ||\omega|| ; \tag{A.16}$$

and the derivatives can be separated by components

$$\frac{\partial \mathbf{q}\left(\left(\omega_k^C + \Omega^C\right)\Delta t\right)}{\partial \omega_k^C} = \begin{pmatrix} \frac{\partial q_0(\omega\Delta t)}{\partial \omega_x} & \frac{\partial q_0(\omega\Delta t)}{\partial \omega_y} & \frac{\partial q_0(\omega\Delta t)}{\partial \omega_z} \\ \frac{\partial q_1(\omega\Delta t)}{\partial \omega_x} & \frac{\partial q_1(\omega\Delta t)}{\partial \omega_y} & \frac{\partial q_1(\omega\Delta t)}{\partial \omega_z} \\ \frac{\partial q_2(\omega\Delta t)}{\partial \omega_x} & \frac{\partial q_2(\omega\Delta t)}{\partial \omega_y} & \frac{\partial q_2(\omega\Delta t)}{\partial \omega_z} \\ \frac{\partial q_3(\omega\Delta t)}{\partial \omega_x} & \frac{\partial q_3(\omega\Delta t)}{\partial \omega_y} & \frac{\partial q_3(\omega\Delta t)}{\partial \omega_z} \end{pmatrix} , \tag{A.17}$$

where each component is computed as follows

$$\left.\frac{\partial q_0(\omega \Delta t)}{\partial \omega_i}\right|_{i=1,\ldots,3} = -\frac{\Delta t}{2}\frac{\omega_i}{\theta}\sin\left(\theta\frac{\Delta t}{2}\right) \tag{A.18}$$

$$\left.\frac{\partial q_i(\omega \Delta t)}{\partial \omega_i}\right|_{i=1,\ldots,3} = \frac{\Delta t}{2}\left(\frac{\omega_i}{\theta}\right)^2\cos\left(\theta\frac{\Delta t}{2}\right) + \frac{1}{\theta}\left(1-\left(\frac{\omega_i}{\theta}\right)^2\right)\sin\left(\theta\frac{\Delta t}{2}\right) \tag{A.19}$$

$$\left.\frac{\partial q_i(\omega \Delta t)}{\partial \omega_j}\right|_{i=1,\ldots,3;i\neq j} = \frac{\Delta t}{2}\frac{\omega_i\omega_j}{\theta^2}\cos\left(\theta\frac{\Delta t}{2}\right) - \frac{1}{\theta}\sin\left(\theta\frac{\Delta t}{2}\right). \tag{A.20}$$

A.2.2 Measurement Model and Derivatives

We will summarize again here the measurement model for completitude before going into its derivatives. For a more elaborated description of this model the reader is referred to section 3.2.

The measurement model can be divided in three steps. In the first of them, features referred to the world reference frame W are converted into Euclidean ones referred to the camera reference frame C. In the case of an inverse depth feature $\mathbf{y}_\rho = (x\ y\ z\ \theta\ \phi\ \rho)^\top$

$$\mathbf{h}_\rho^C = \mathbf{R}^{CW}\left(\rho\left(\begin{pmatrix} x \\ y \\ z \end{pmatrix} - \mathbf{r}^{WC}\right) + \mathbf{m}(\theta,\phi)\right); \tag{A.21}$$

and for a feature coded in Euclidean form $\mathbf{y}_{\mathrm{XYZ}} = (X\ Y\ Z)^\top$

$$\mathbf{h}_{\mathrm{XYZ}}^C = \mathbf{R}^{CW}\left(\begin{pmatrix} X \\ Y \\ Z \end{pmatrix} - \mathbf{r}^{WC}\right). \tag{A.22}$$

A pinhole camera model is then applied

$$\mathbf{h}_u = \begin{pmatrix} u_u \\ v_u \end{pmatrix} = \begin{pmatrix} C_x - \frac{f}{d_x}\frac{h_x^C}{h_z^C} \\ C_y - \frac{f}{d_y}\frac{h_y^C}{h_z^C} \end{pmatrix}, \tag{A.23}$$

that gives us the 2D image coordinates $\mathbf{h}_u$ assuming a pure projective model. In order to cope with the distortions coming from real lenses, we add a radial distortion model to the ideal undistorted coordinates. In this book we have used the standard two parameter distortion model from photogrammetry [Mikhail *et al.* 2001], which is described next.

The ideal projective undistorted coordinates $\mathbf{h}_u = (u_u, v_u)^\top$ are recovered from the real distorted ones $\mathbf{h}_d = (u_d, v_d)^\top$ as follows,

$$\mathbf{h}_u = \begin{pmatrix} C_x + (u_d - C_x)\left(1 + \kappa_1 r_d^2 + \kappa_2 r_d^4\right) \\ C_y + (v_d - C_y)\left(1 + \kappa_1 r_d^2 + \kappa_2 r_d^4\right) \end{pmatrix}$$

$$r_d = \sqrt{(d_x (u_d - C_x))^2 + (d_y (v_d - C_y))^2} \tag{A.24}$$

Where $(C_x \ C_y)^\top$ are the principal point coordinates; and κ_1 and κ_2 the radial distortion coefficients.

To compute the distorted coordinates from the undistorted:

$$\mathbf{h}_d = \begin{pmatrix} C_x + \frac{(u_u - C_x)}{\left(1+\kappa_1 r_d^2 + \kappa_2 r_d^4\right)} \\ C_y + \frac{(v_u - C_y)}{\left(1+\kappa_1 r_d^2 + \kappa_2 r_d^4\right)} \end{pmatrix} \tag{A.25}$$

$$r_u = r_d\left(1 + \kappa_1 r_d^2 + \kappa_2 r_d^4\right) \tag{A.26}$$

$$r_u = \sqrt{(d_x (u_u - C_x))^2 + (d_y (v_u - C_y))^2} \tag{A.27}$$

r_u is readily computed computed from (A.27), but r_d has to be numerically solved from (A.26), e.g using Newton-Raphson, hence (A.25) can be used to compute the distorted point.

The Jacobian of this measurement equation by the state vector is extracted from the model defined above. The full Jacobian $\mathbf{H}$ is divided into rows, each one corresponding to a point measurement

$$\mathbf{H} = \begin{pmatrix} \frac{\partial \mathbf{h}_1}{\partial \mathbf{x}} \\ \vdots \\ \frac{\partial \mathbf{h}_i}{\partial \mathbf{x}} \\ \vdots \\ \frac{\partial \mathbf{h}_m}{\partial \mathbf{x}} \end{pmatrix}. \tag{A.28}$$

The derivative of each measurement $\frac{\partial \mathbf{h}_i}{\partial \mathbf{x}}$ can be separated into derivatives by the camera states $\frac{\partial \mathbf{h}_i}{\partial \mathbf{x}_C}$ and derivatives by the map features $\frac{\partial \mathbf{h}_i}{\partial \mathbf{x}_M}$

$$\frac{\partial \mathbf{h}_i}{\partial \mathbf{x}} = \left(\frac{\partial \mathbf{h}_i}{\partial \mathbf{x}_C} \frac{\partial \mathbf{h}_i}{\partial \mathbf{x}_M} \right). \tag{A.29}$$

As the camera state vector is composed by its location $\mathbf{r}^{WC}$, orientation $\mathbf{q}^{WC}$ and linear and angular velocities $\mathbf{v}$ and ω (equation A.8); the partial derivative $\frac{\partial \mathbf{h}_i}{\partial \mathbf{x}_C}$ can also be further divided

$$\frac{\partial \mathbf{h}_i}{\partial \mathbf{x}_C} = \begin{pmatrix} \frac{\partial \mathbf{h}_i}{\partial \mathbf{r}^{WC}} & \frac{\partial \mathbf{h}_i}{\partial \mathbf{q}^{WC}} & \cancelto{0}{\frac{\partial \mathbf{h}_i}{\partial \mathbf{v}}} & \cancelto{0}{\frac{\partial \mathbf{h}_i}{\partial \omega}} \end{pmatrix} \tag{A.30}$$

As camera velocities are not involved in the measurement model, the derivatives involving velocities $\frac{\partial \mathbf{h}_i}{\partial \mathbf{v}}$ and $\frac{\partial \mathbf{h}_i}{\partial \omega}$ cancel out to zero. The chain rule will be used to compute the rest of the partial derivatives. Starting by $\frac{\partial \mathbf{h}_i}{\partial \mathbf{r}^{WC}}$,

$$\frac{\partial \mathbf{h}_i}{\partial \mathbf{r}^{WC}} = \frac{\partial \mathbf{h}_d}{\partial \mathbf{h}_u} \frac{\partial \mathbf{h}_u}{\partial \mathbf{h}^C} \frac{\partial \mathbf{h}^C}{\partial \mathbf{r}^{WC}} \tag{A.31}$$

We have to compute first the partial derivative $\frac{\partial \mathbf{h}_d}{\partial \mathbf{h}_u}$ from equation A.25. As this equation is not in explicit form, we will calculate first the Jacobian for the undistortion $\frac{\partial \mathbf{h}_u}{\partial \mathbf{h}_d}$ from formula A.57:

$$\frac{\partial \mathbf{h}_u}{\partial \mathbf{h}_d} = \left(\begin{array}{c|c} \begin{array}{c} \left(1+\kappa_1 r_d^2+\kappa_2 r_d^4\right)+ \\ 2\left(\left(u_d-C_x\right)d_x\right)^2\left(\kappa_1+2\kappa_2 r_d^2\right) \end{array} & \begin{array}{c} 2d_y^2\left(u_d-C_x\right)\left(v_d-C_y\right)\cdot \\ \cdot\left(\kappa_1+2\kappa_2 r_d^2\right) \end{array} \\ \hline \begin{array}{c} 2d_x^2\left(v_d-C_y\right)\left(u_d-C_x\right)\cdot \\ \cdot\left(\kappa_1+2\kappa_2 r_d^2\right) \end{array} & \begin{array}{c} \left(1+\kappa_1 r_d^2+\kappa_2 r_d^4\right)+ \\ 2\left(\left(v_d-C_y\right)d_y\right)^2\left(\kappa_1+2\kappa_2 r_d^2\right) \end{array} \end{array}\right) \tag{A.32}$$

The Jacobian for the distortion is computed by inverting the previous matrix $\frac{\partial \mathbf{h}_u}{\partial \mathbf{h}_d}$ (A.32):

$$\frac{\partial \mathbf{h}_d}{\partial \mathbf{h}_u} = \left(\frac{\partial \mathbf{h}_u}{\partial \mathbf{h}_d}\right)^{-1} \tag{A.33}$$

The derivatives for the pinhole camera model are easily extracted from A.23

$$\frac{\partial \mathbf{h}_u}{\partial \mathbf{h}^C} = \begin{pmatrix} \frac{-f}{d_x h_z^C} & 0 & \frac{h_x^C f}{d_x h_z^{C2}} \\ 0 & \frac{-f}{d_y h_z^C} & \frac{h_y^C f}{d_y h_z^{C2}} \end{pmatrix} \tag{A.34}$$

And finally, the partial derivative $\frac{\partial \mathbf{h}^C}{\partial \mathbf{r}^{WC}}$ is

$$\frac{\partial \mathbf{h}_\rho^C}{\partial \mathbf{r}^{WC}} = -\rho \mathbf{R}^{CW} \tag{A.35}$$

$$\frac{\partial \mathbf{h}^C{}_{\mathrm{XYZ}}}{\partial \mathbf{r}^{WC}} = -\mathbf{R}^{CW}\ ; \tag{A.36}$$

for the cases of an inverse depth feature and a Euclidean XYZ one respectively.

The partial derivative of the measurement function $\mathbf{h}_i$ by the quaternion rotation $\mathbf{q}^{WC}$ is again computed using the chain rule

$$\frac{\partial \mathbf{h}_i}{\partial \mathbf{q}^{WC}} = \frac{\partial \mathbf{h}_d}{\partial \mathbf{h}_u}\frac{\partial \mathbf{h}_u}{\partial \mathbf{h}^C}\frac{\partial \mathbf{h}^C}{\partial \mathbf{q}^{WC}} \tag{A.37}$$

Partial derivatives $\frac{\partial \mathbf{h}_d}{\partial \mathbf{h}_u}$ and $\frac{\partial \mathbf{h}_u}{\partial \mathbf{h}^C}$ have already been detailed in formulas A.33 and A.34 respectively. The only calculation left is $\frac{\partial \mathbf{h}^C}{\partial \mathbf{q}^{WC}}$. Such computation can be done dividing first the jacobian into two pieces

$$\frac{\partial \mathbf{h}^C}{\partial \mathbf{q}^{WC}} = \frac{\partial \mathbf{h}^C}{\partial \mathbf{q}^{CW}}\frac{\partial \mathbf{q}^{CW}}{\partial \mathbf{q}^{WC}}\ ; \tag{A.38}$$

where

$$\frac{\partial \mathbf{q}^{CW}}{\partial \mathbf{q}^{WC}} = \begin{pmatrix} 1 & 0 & 0 & 0 \\ 0 & -1 & 0 & 0 \\ 0 & 0 & -1 & 0 \\ 0 & 0 & 0 & -1 \end{pmatrix}. \tag{A.39}$$

Partial derivatives $\frac{\partial \mathbf{h}^C}{\partial \mathbf{q}^{CW}}$ can be computed dividing them as the derivative with respect to each one of the components of the quaternion $\mathbf{q}^{CW} = \left(q_0^{CW}\ q_1^{CW}\ q_2^{CW}\ q_3^{CW}\right)^\top$,

$$\frac{\partial \mathbf{h}^C}{\partial \mathbf{q}^{CW}} = \left(\frac{\partial \mathbf{h}^C}{\partial q_0^{CW}}\ \frac{\partial \mathbf{h}^C}{\partial q_1^{CW}}\ \frac{\partial \mathbf{h}^C}{\partial q_2^{CW}}\ \frac{\partial \mathbf{h}^C}{\partial q_3^{CW}} \right)\ ; \tag{A.40}$$

where the derivative with respect to the quaternion components only affects to the rotation matrix

$$\frac{\partial \mathbf{h}^C}{\partial q_0^{CW}} = \frac{\partial \mathbf{R}^{CW}}{\partial q_0^{CW}} \left(\rho_i \left(\begin{pmatrix} x_i \\ y_i \\ z_i \end{pmatrix} - \mathbf{r}^{WC} \right) + \mathbf{m}(\theta_i, \phi_i) \right) \tag{A.41}$$

$$\frac{\partial \mathbf{h}^C}{\partial q_1^{CW}} = \frac{\partial \mathbf{R}^{CW}}{\partial q_1^{CW}} \left(\rho_i \left(\begin{pmatrix} x_i \\ y_i \\ z_i \end{pmatrix} - \mathbf{r}^{WC} \right) + \mathbf{m}(\theta_i, \phi_i) \right) \tag{A.42}$$

$$\frac{\partial \mathbf{h}^C}{\partial q_2^{CW}} = \frac{\partial \mathbf{R}^{CW}}{\partial q_2^{CW}} \left(\rho_i \left(\begin{pmatrix} x_i \\ y_i \\ z_i \end{pmatrix} - \mathbf{r}^{WC} \right) + \mathbf{m}(\theta_i, \phi_i) \right) \tag{A.43}$$

$$\frac{\partial \mathbf{h}^C}{\partial q_3^{CW}} = \frac{\partial \mathbf{R}^{CW}}{\partial q_3^{CW}} \left(\rho_i \left(\begin{pmatrix} x_i \\ y_i \\ z_i \end{pmatrix} - \mathbf{r}^{WC} \right) + \mathbf{m}(\theta_i, \phi_i) \right). \tag{A.44}$$

From the conversion formula from quaternion to rotation matrix orientation representation,

$$\mathbf{R} = \begin{pmatrix} q_0^2+q_1^2-q_2^2-q_3^2 & 2(q_1q_2-q_0q_3) & 2(q_3q_1+q_0q_2) \\ 2(q_1q_2+q_0q_3) & q_0^2-q_1^2+q_2^2-q_3^2 & 2(q_2q_3-q_0q_1) \\ 2(q_3q_1-q_0q_2) & 2(q_2q_3+q_0q_1) & q_0^2-q_1^2-q_2^2+q_3^2 \end{pmatrix} . \tag{A.45}$$

The derivatives of rotation matrix $\mathbf{R}^{CW}$ with respect to each component of the quaternion are easily computed

$$\frac{\partial \mathbf{R}^{CW}}{\partial q_0^{CW}} = \begin{pmatrix} 2q_0^{CW} & -2q_3^{CW} & 2q_2^{CW} \\ 2q_3^{CW} & 2q_0^{CW} & -2q_1^{CW} \\ -2q_2^{CW} & 2q_1^{CW} & 2q_0^{CW} \end{pmatrix} \tag{A.46}$$

$$\frac{\partial \mathbf{R}^{CW}}{\partial q_1^{CW}} = \begin{pmatrix} 2q_1^{CW} & 2q_2^{CW} & 2q_3^{CW} \\ 2q_2^{CW} & -2q_1^{CW} & -2q_0^{CW} \\ 2q_3^{CW} & 2q_0^{CW} & -2q_1^{CW} \end{pmatrix} \tag{A.47}$$

$$\frac{\partial \mathbf{R}^{CW}}{\partial q_2^{CW}} = \begin{pmatrix} -2q_2^{CW} & 2q_1^{CW} & 2q_0^{CW} \\ 2q_1^{CW} & 2q_2^{CW} & 2q_3^{CW} \\ -2q_0^{CW} & 2q_3^{CW} & -2q_2^{CW} \end{pmatrix} \tag{A.48}$$

$$\frac{\partial \mathbf{R}^{CW}}{\partial q_3^{CW}} = \begin{pmatrix} -2q_3^{CW} & -2q_0^{CW} & 2q_1^{CW} \\ 2q_0^{CW} & -2q_3^{CW} & 2q_2^{CW} \\ 2q_1^{CW} & 2q_2^{CW} & 2q_3^{CW} \end{pmatrix} \tag{A.49}$$

In the derivative of each measurement by the map $\frac{\partial \mathbf{h}_i}{\partial x_M}$ appearing in equation A.29 only the feature i is in the measurement equation, so

$$\frac{\partial \mathbf{h}_i}{\partial x_M} = \left(0 \ldots 0 \frac{\partial \mathbf{h}_i}{\partial \mathbf{y}_i} 0 \ldots 0 \right) . \tag{A.50}$$

The partial derivative $\frac{\partial \mathbf{h}_i}{\partial \mathbf{y}_i}$ is as follows

$$\frac{\partial \mathbf{h}_i}{\partial \mathbf{y}_i} = \frac{\partial \mathbf{h}_d}{\partial \mathbf{h}_u} \frac{\partial \mathbf{h}_u}{\partial \mathbf{h}^C} \frac{\partial \mathbf{h}^C}{\partial \mathbf{y}_i} ; \tag{A.51}$$

where $\frac{\partial \mathbf{h}_d}{\partial \mathbf{h}_u}$ and $\frac{\partial \mathbf{h}_u}{\partial \mathbf{h}^C}$ are detailed in A.33 and A.34. The partial derivative $\frac{\partial \mathbf{h}^C}{\partial \mathbf{y}_i}$ in the above product is, for the inverse depth case

$$\frac{\partial \mathbf{h}_\rho^C}{\partial \mathbf{y}_i} = \left(\rho \mathbf{R}^{CW} | \; \mathbf{R}^{CW} \frac{\partial \mathbf{m}}{\partial \theta} | \; \mathbf{R}^{CW} \frac{\partial \mathbf{m}}{\partial \phi} | \; \mathbf{R}^{CW} \left(\begin{pmatrix} x_i \\ y_i \\ z_i \end{pmatrix} - \mathbf{r}^{WC} \right) \right) . \tag{A.52}$$

Here $\mathbf{m} = \begin{pmatrix} \cos\phi \sin\theta & -\sin\phi & \cos\phi\cos\theta \end{pmatrix}^\top$ is the unit vector defined by the azimuth-elevation pair $(\theta \;, \; \phi)$, derivatives $\frac{\partial \mathbf{m}}{\partial \theta}$ and $\frac{\partial \mathbf{m}}{\partial \phi}$ come straightforwardly as

$$\frac{\partial \mathbf{m}}{\partial \theta} = (cos\phi cos\theta \;\; 0 \;\; -cos\phi sin\theta)^{\top} \tag{A.53}$$

$$\frac{\partial \mathbf{m}}{\partial \phi} = (-sin\phi sin\theta \;\; -cos\phi \;\; -sin\phi cos\theta)^{\top} \; . \tag{A.54}$$

For a XYZ feature, this latest partial derivative is

$$\frac{\partial \mathbf{h}^{C}_{\mathrm{XYZ}}}{\partial \mathbf{y}_i} = \mathbf{R}^{CW} \; . \tag{A.55}$$

A.2.3 Inverse Depth Point Feature Initialization and Derivatives

The initialization function defines a new point feature $\mathbf{y}^{NEW}$ from an image point $\mathbf{h}$, the current state vector $\mathbf{x}$ and and initial value for the inverse depth ρ_0

$$\mathbf{y}^{NEW} = \mathbf{y}(\mathbf{x}, \mathbf{h}, \rho_0) \tag{A.56}$$

The initialization function follows the same three steps than the measurement equation but in reverse order. First, the image point has to be undistorted

$$\mathbf{h}_u = \begin{pmatrix} C_x + (u_d - C_x)\left(1 + \kappa_1 r_d^2 + \kappa_2 r_d^4\right) \\ C_y + (v_d - C_y)\left(1 + \kappa_1 r_d^2 + \kappa_2 r_d^4\right) \end{pmatrix}$$

$$r_d = \sqrt{\left(d_x\left(u_d - C_x\right)\right)^2 + \left(d_y\left(v_d - C_y\right)\right)^2} \tag{A.57}$$

The 3D ray –in the camera reference frame– where the point feature lies can be extracted from the line joining the optical centre and the undistorted image coordinates

$$\mathbf{h}^C = \begin{pmatrix} \frac{-(u_u - C_x)d_x}{f} \\ \frac{-(v_u - C_y)d_y}{f} \\ 1 \end{pmatrix} . \tag{A.58}$$

Using the current camera orientation estimation from the state vector, this ray can be transformed to the world reference frame and the azimuth and elevation angles extracted;

$$\mathbf{h}^W = \mathbf{R}^{WC}\left(\mathbf{q}^{WC}\right)\mathbf{h}^C \; , \tag{A.59}$$

$$\theta = arctan\left(h_x^W, h_z^W\right) \; , \tag{A.60}$$

$$\phi = arctan\left(-h_y^W, \sqrt{{h_x^W}^2 + {h_z^W}^2}\right) \; . \tag{A.61}$$

The position of the optical centre is directly extracted from the current camera position,

$$\begin{pmatrix} x_i \\ y_i \\ z_i \end{pmatrix} = \mathbf{r}^{WC} . \tag{A.62}$$

The initial value for the inverse depth ρ_0 is chosen to cover in its 95% acceptance region a range of possible depths covering from a minimum close distance d_{min} to infinity. A typical value in our experiments is $\rho_0 = 1$ and $\sigma_{\rho_0} = 1$; covering a range from $d_{min} = 0.33$ to infinite (and beyond) with a probability of 0.95.

The newly initialized feature $\mathbf{y}^{NEW} = (x_i \; y_i \; z_i \; \theta \; \phi \; \rho_0)^\top$ is added to the state vector

$$\mathbf{x}^{NEW} = \begin{pmatrix} \mathbf{x}^{OLD} \\ \mathbf{y}^{NEW} \end{pmatrix} , \tag{A.63}$$

And the state covariance is updated in the following manner

$$\mathbf{P}^{NEW} = \mathbf{J} \begin{pmatrix} \mathbf{P}^{OLD} & 0 & 0 \\ 0 & \mathbf{R} & 0 \\ 0 & 0 & \sigma_{\rho_0} \end{pmatrix} \mathbf{J}^\top ; \tag{A.64}$$

being $\mathbf{R}$ the image noise covariance associated with our feature detector.

The matrix $\mathbf{J}$ is the Jacobian of this initialization function

$$\mathbf{J} = \begin{pmatrix} & \mathbf{I} & & \begin{matrix} 0 \\ \vdots \\ 0 \end{matrix} \\ \frac{\partial \mathbf{y}}{\partial \mathbf{x}_C} & & 0 \ldots 0 & \frac{\partial \mathbf{y}}{\partial \mathbf{h}} \end{pmatrix} . \tag{A.65}$$

As it happened in previous section, point un-projection does not depend either on the camera velocities, so the partial derivatives by the camera parameters has non-zero terms in camera position and orientation

$$\frac{\partial \mathbf{y}}{\partial \mathbf{x}_C} = \begin{pmatrix} \frac{\partial \mathbf{y}}{\partial \mathbf{r}^{WC}} & \frac{\partial \mathbf{y}}{\partial \mathbf{q}^{WC}} & \cancelto{0}{\frac{\partial \mathbf{y}}{\partial \mathbf{v}}} & \cancelto{0}{\frac{\partial \mathbf{y}}{\partial \omega}} \end{pmatrix} . \tag{A.66}$$

The Jacobian by the camera position is

$$\frac{\partial \mathbf{y}}{\partial \mathbf{r}^{WC}} = \begin{pmatrix} \mathbf{I} \\ \mathbf{0} \end{pmatrix} . \tag{A.67}$$

The derivatives by the orientation quaternion have non-zero terms in the azimuth and elevation angles' positions:

$$\frac{\partial \mathbf{y}}{\partial \mathbf{q}^{WC}} = \begin{pmatrix} \mathbf{0} \\ \frac{\partial \theta}{\partial \mathbf{q}^{WC}} \\ \frac{\partial \phi}{\partial \mathbf{q}^{WC}} c \\ 0 \end{pmatrix} ; \tag{A.68}$$

where both derivatives $\frac{\partial \theta}{\partial \mathbf{q}^{WC}}$ and $\frac{\partial \phi}{\partial \mathbf{q}^{WC}}$ are as follows

$$\frac{\partial \theta}{\partial \mathbf{q}^{WC}} = \frac{\partial \theta}{\partial \mathbf{h}^W} \frac{\partial \mathbf{h}^W}{\partial \mathbf{q}^{WC}} \tag{A.69}$$

$$\frac{\partial \phi}{\partial \mathbf{q}^{WC}} = \frac{\partial \phi}{\partial \mathbf{h}^W} \frac{\partial \mathbf{h}^W}{\partial \mathbf{q}^{WC}} ; \tag{A.70}$$

$$\frac{\partial \theta}{\partial \mathbf{h}^W} = \begin{pmatrix} \frac{\mathbf{h}_z^W}{\mathbf{h}_x^{W2} + \mathbf{h}_z^{W2}} & 0 & -\frac{\mathbf{h}_x^W}{\mathbf{h}_x^{W2} + \mathbf{h}_z^{W2}} \end{pmatrix} \tag{A.71}$$

$$\frac{\partial \phi}{\partial \mathbf{h}^W} = \begin{pmatrix} \frac{\mathbf{h}_x^W \mathbf{h}_y^W}{(\mathbf{h}_x^{W2} + \mathbf{h}_y^{W2} + \mathbf{h}_z^{W2})\sqrt{\mathbf{h}_x^{W2} + \mathbf{h}_z^{W2}}} \\ -\frac{\sqrt{\mathbf{h}_x^{W2} + \mathbf{h}_z^{W2}}}{\mathbf{h}_x^{W2} + \mathbf{h}_y^{W2} + \mathbf{h}_z^{W2}} \\ \frac{\mathbf{h}_z^W \mathbf{h}_y^W}{(\mathbf{h}_x^{W2} + \mathbf{h}_y^{W2} + \mathbf{h}_z^{W2})\sqrt{\mathbf{h}_x^{W2} + \mathbf{h}_z^{W2}}} \end{pmatrix}^\top \tag{A.72}$$

$$\frac{\partial \mathbf{h}^W}{\partial \mathbf{q}^{WC}} = \begin{pmatrix} \frac{\partial \mathbf{h}^W}{\partial \mathbf{q}_0^{WC}} & \frac{\partial \mathbf{h}^W}{\partial \mathbf{q}_1^{WC}} & \frac{\partial \mathbf{h}^W}{\partial \mathbf{q}_2^{WC}} & \frac{\partial \mathbf{h}^W}{\partial \mathbf{q}_3^{WC}} \end{pmatrix} \tag{A.73}$$

$$\frac{\partial \mathbf{h}^W}{\partial \mathbf{q}_i^{WC}} = \mathbf{h}^C \frac{\partial \mathbf{R}^{WC}}{\partial \mathbf{q}_i^{WC}} . \tag{A.74}$$

The derivatives of the rotation matrix $\mathbf{R}^{WC}$ by each quaternion component $\mathbf{q}_i^{WC}$ have been detailed in equations A.46 to A.49.

Finally, the derivatives by the salient image point $\mathbf{h}$

$$\frac{\partial \mathbf{y}}{\partial \mathbf{h}} = \begin{pmatrix} \frac{\partial \mathbf{y}\prime}{\partial \mathbf{h}} & 0 \\ 0 & 1 \end{pmatrix} , \tag{A.75}$$

where $\mathbf{y}\prime = (x_i \ y_i \ z_i \ \theta_i \ \phi_i)$ stands for the feature parameters computed from the salient image point $\mathbf{h}$. That is, all of them except the inverse depth one ρ_0. The derivative $\frac{\partial \mathbf{y}\prime}{\partial \mathbf{h}}$ is computed as

$$\frac{\partial \mathbf{y}'}{\partial \mathbf{h}} = \frac{\partial \mathbf{y}'}{\partial \mathbf{h}^W} \frac{\partial \mathbf{h}^W}{\partial \mathbf{h}^C} \frac{\partial \mathbf{h}^C}{\partial \mathbf{h}_u} \frac{\partial \mathbf{h}_u}{\partial \mathbf{h}_d} ; \tag{A.76}$$

where

$$\frac{\partial \mathbf{y}'}{\partial \mathbf{h}^W} = \begin{pmatrix} \mathbf{0} \frac{\partial \theta}{\partial \mathbf{h}^W} \frac{\partial \phi}{\partial \mathbf{h}^W} \end{pmatrix} \quad \text{(A.77)}$$

with $\frac{\partial \theta}{\partial \mathbf{h}^W}$ and $\frac{\partial \phi}{\partial \mathbf{h}^W}$ being already computed in equations A.71 and A.72; and

$$\frac{\partial \mathbf{h}^W}{\partial \mathbf{h}^C} = \mathbf{R}^{WC} ; \quad \text{(A.78)}$$

$$\frac{\partial \mathbf{h}^C}{\partial \mathbf{h}_u} = \begin{pmatrix} \frac{d_x}{f} & 0 & 0 \\ 0 & \frac{d_x}{f} & 0 \end{pmatrix} . \quad \text{(A.79)}$$

The Jacobian of the undistortion is detailed in equation A.32.

A.3 Uncalibrated EKF-Based SfM

A.3.1 Dynamic Model and Derivatives

In the uncalibrated case, the camera state vector $\mathbf{x}_C$ is augmented with the projective parameters –focal length f and principal point coordinates C_x and C_y– and the parameters modeling the lens distortion –in our case, κ_1 and κ_2 for the radial distortion

$$\mathbf{x}_C = \begin{pmatrix} f \\ C_x \\ C_y \\ \kappa_1 \\ \kappa_2 \\ \mathbf{r}^{WC} \\ \mathbf{q}^{WC} \\ \mathbf{v}^W \\ \omega^C \end{pmatrix} \quad \text{(A.80)}$$

The motion model assumed for the camera is the same as detailed in section A.2.1. The calibration parameters are assumed to remain constant in the experiments of the book as we did not deal with zooming cameras. Nevertheless, the addition of a dynamic model for the camera calibration parameters would be straightforward. The dynamic model for the camera state vector is

$$\mathbf{x}_{C_{k+1}} = \begin{pmatrix} f_{k+1} \\ C_{xk+1} \\ C_{yk+1} \\ \kappa_{1k+1} \\ \kappa_{2k+1} \\ \mathbf{r}^{WC}_{k+1} \\ \mathbf{q}^{WC}_{k+1} \\ \mathbf{v}^{W}_{k+1} \\ \omega^{C}_{k+1} \end{pmatrix} = \mathbf{f}_v\left(\mathbf{x}_{C_k}, \mathbf{n}\right) = \begin{pmatrix} f_k \\ C_{xk} \\ C_{yk} \\ \kappa_{1k} \\ \kappa_{2k} \\ \mathbf{r}^{WC}_{k} + \left(\mathbf{v}^{W}_{k} + \mathbf{V}^{W}_{k}\right)\Delta t \\ \mathbf{q}^{WC}_{k} \times \mathbf{q}\left(\left(\omega^{C}_{k} + \Omega^{C}\right)\Delta t\right) \\ \mathbf{v}^{W}_{k} + \mathbf{V}^{W} \\ \omega^{C}_{k} + \Omega^{C} \end{pmatrix}. \tag{A.81}$$

And the derivatives for the uncalibrated case can be easily extracted from the ones in section A.2.1

$$\frac{\partial \mathbf{f}_v}{\partial \mathbf{x}_v} = \begin{pmatrix} I & \mathbf{0} & \mathbf{0} & \mathbf{0} & \mathbf{0} \\ \mathbf{0} & I & \mathbf{0} & \Delta t I & \mathbf{0} \\ \mathbf{0} & \mathbf{0} & \frac{\partial \mathbf{q}^{WC}_{k+1}}{\partial \mathbf{q}^{WC}_{k}} & \mathbf{0} & \frac{\partial \mathbf{q}^{WC}_{k+1}}{\partial \omega^{C}_{k}} \\ \mathbf{0} & \mathbf{0} & \mathbf{0} & I & \mathbf{0} \\ \mathbf{0} & \mathbf{0} & \mathbf{0} & \mathbf{0} & I \end{pmatrix} \tag{A.82}$$

$$\frac{\partial \mathbf{f}_v}{\partial \mathbf{n}} = \begin{pmatrix} \mathbf{0} & \mathbf{0} \\ \Delta t I & \mathbf{0} \\ \mathbf{0} & \frac{\partial \mathbf{q}^{WC}_{k+1}}{\partial \Omega^{C}} \\ I & \mathbf{0} \\ \mathbf{0} & I \end{pmatrix} \tag{A.83}$$

A.3.2 Measurement Model and Derivatives

The pinhole camera model plus a two parameter radial distortion model for the lens is still used in the uncalibrated case, so equations A.21 to A.27 in section A.2.2 keep describing the measurement model. The key difference is that, in this case, the calibration parameters are not considered constant but estimated in a joint state vector. Then, the Jacobians are modified with respect to the calibrated case.

The whole Jacobian $\mathbf{H}$ is as follows

$$\mathbf{H} = \begin{pmatrix} \frac{\partial \mathbf{h}_1}{\partial \mathbf{x}} \\ \vdots \\ \frac{\partial \mathbf{h}_i}{\partial \mathbf{x}} \\ \vdots \\ \frac{\partial \mathbf{h}_m}{\partial \mathbf{x}} \end{pmatrix} \tag{A.84}$$

$$\frac{\partial \mathbf{h}_i}{\partial \mathbf{x}} = \begin{pmatrix} \frac{\partial \mathbf{h}_i}{\partial \mathbf{x}_C} \frac{\partial \mathbf{h}_i}{\partial \mathbf{x}_M} \end{pmatrix} , \tag{A.85}$$

the derivative by the camera state $\frac{\partial \mathbf{h}_i}{\partial \mathbf{x}_C}$ is the one that changes with respect to the calibrated case, as now we have to include the derivatives by the calibration parameters

$$\frac{\partial \mathbf{h}_i}{\partial x_C} = \begin{pmatrix} \frac{\partial \mathbf{h}_i}{\partial f} \frac{\partial \mathbf{h}_i}{\partial C_x} \frac{\partial \mathbf{h}_i}{\partial C_y} \frac{\partial \mathbf{h}_i}{\partial \kappa_1} \frac{\partial \mathbf{h}_i}{\partial \kappa_2} \frac{\partial \mathbf{h}_i}{\partial \mathbf{r}^{WC}} \frac{\partial \mathbf{h}_i}{\partial \mathbf{q}^{WC}} \cancelto{0}{\frac{\partial \mathbf{h}_i}{\partial \mathbf{v}}} \cancelto{0}{\frac{\partial \mathbf{h}_i}{\partial \omega}} \end{pmatrix} . \tag{A.86}$$

The computation of the aboves derivatives can be simplified by slightly modifying the pinhole camera model equations in section A.2.2. The transformation from the world frame to the camera frame in equations A.21 and A.22 is the same, but from there we introduce the intermediate variables $\mathbf{h}_u^*$ and $\mathbf{h}_d^*$, that represents the image projection of the 3D point in a reference frame anchored in the principal point

$$\mathbf{h}_u^* = \begin{pmatrix} u_u^* \\ v_u^* \end{pmatrix} = \begin{pmatrix} \frac{-f}{d_x} \frac{h_x^C}{h_z^C} \\ \frac{-f}{d_y} \frac{h_y^C}{h_z^C} \end{pmatrix} \tag{A.87}$$

$$\mathbf{h}_u^* = \begin{pmatrix} u_u^* \\ v_u^* \end{pmatrix} = \mathbf{h}_d^* \left(1 + \kappa_1 r_d^2 + \kappa_2 r_d^4\right) \tag{A.88}$$

$$r_d = \sqrt{\left(d_x u_d^*\right)^2 + \left(d_y v_d^*\right)^2} . \tag{A.89}$$

The image reference can be moved by adding the optical center coordinates to the "starred" image coordinates

$$\mathbf{h}_u = \mathbf{h}_u^* + \begin{pmatrix} C_x \\ C_y \end{pmatrix} \tag{A.90}$$

$$\mathbf{h}_d = \mathbf{h}_d^* + \begin{pmatrix} C_x \\ C_y \end{pmatrix} . \tag{A.91}$$

The derivatives with respect to the projective parameters $\frac{\partial \mathbf{h}_i}{\partial f}$, $\frac{\partial \mathbf{h}_i}{\partial C_x}$ and $\frac{\partial \mathbf{h}_i}{\partial C_y}$ can be straightforwardly extracted from equations A.87 to A.91:

$$\frac{\partial \mathbf{h}_d}{\partial f} = \frac{\partial \mathbf{h}_d}{\partial \mathbf{h}_d^*} \frac{\partial \mathbf{h}_d^*}{\partial \mathbf{h}_u^*} \frac{\partial \mathbf{h}_u^*}{\partial f} \tag{A.92}$$

$$\frac{\partial \mathbf{h}_d}{\partial \mathbf{h}_d^*} = \mathbf{I} \tag{A.93}$$

$$\frac{\partial \mathbf{h}_u^*}{\partial f} = \begin{pmatrix} \frac{-h_x^C}{d_x h_z^C} \\ \frac{-h_y^C}{d_y h_z^C} \end{pmatrix} \tag{A.94}$$

$$\frac{\partial \mathbf{h}_d}{\partial C_x} = \begin{pmatrix} 1 \\ 0 \end{pmatrix} \tag{A.95}$$

$$\frac{\partial \mathbf{h}_d}{\partial C_y} = \begin{pmatrix} 0 \\ 1 \end{pmatrix} . \tag{A.96}$$

The derivatives by the radial distortion parameters $\frac{\partial \mathbf{h}_d}{\partial \kappa_1}$ and $\frac{\partial \mathbf{h}_d}{\partial \kappa_2}$ are a bit more involved. Starting from equation A.88, the derivatives of the undistorted image coordinates $\mathbf{h}_u^*$ can be easily computed:

$$\frac{\partial u_u^*}{\partial \kappa_1} = u_d^* r_d^2 \tag{A.97}$$

$$\frac{\partial v_u^*}{\partial \kappa_1} = v_d^* r_d^2 \tag{A.98}$$

$$\frac{\partial u_u^*}{\partial \kappa_2} = u_d^* r_d^4 \tag{A.99}$$

$$\frac{\partial v_u^*}{\partial \kappa_2} = v_d^* r_d^4 . \tag{A.100}$$

Doing implicit differentiation on equation A.88, we have the following

$$\frac{\partial \mathbf{h}_u^*}{\partial \kappa_1} = \frac{\partial \mathbf{h}_d^*}{\partial \kappa_1} \left(1 + \kappa_1 r_d^2 + \kappa_2 r_d^4\right) + \mathbf{h}_d^* \frac{\partial \left(1 + \kappa_1 r_d^2 + \kappa_2 r_d^4\right)}{\partial \kappa_1} . \tag{A.101}$$

Being $r_d = \sqrt{u_d^{*2} + v_d^{*2}}$, we can expand $\frac{\partial \left(1+\kappa_1 r_d^2+\kappa_2 r_d^4\right)}{\partial \kappa_1}$ as follows

$$\frac{\partial \left(1 + \kappa_1 r_d^2 + \kappa_2 r_d^4\right)}{\partial \kappa_1} = r_d^2 + 2\kappa_1 \left(\frac{\partial u_d^*}{\kappa_1} + \frac{\partial v_d^*}{\kappa_1}\right) + 4\kappa_2 \left(\frac{\partial u_d^*}{\kappa_1} + \frac{\partial v_d^*}{\kappa_1}\right) \tag{A.102}$$

Doing the same with the derivative by κ_2

$$\frac{\partial \mathbf{h}_u^*}{\partial \kappa_2} = \frac{\partial \mathbf{h}_d^*}{\partial \kappa_2}\left(1+\kappa_1 r_d^2+\kappa_2 r_d^4\right) + \mathbf{h}_d^* \frac{\partial\left(1+\kappa_1 r_d^2+\kappa_2 r_d^4\right)}{\partial \kappa_2} \tag{A.103}$$

$$\frac{\partial\left(1+\kappa_1 r_d^2+\kappa_2 r_d^4\right)}{\partial \kappa_2} = r_d^4 + 2\kappa_1\left(\frac{\partial u_d^*}{\kappa_2}+\frac{\partial v_d^*}{\kappa_2}\right) + 4\kappa_2\left(\frac{\partial u_d^*}{\kappa_2}+\frac{\partial v_d^*}{\kappa_2}\right). \tag{A.104}$$

From A.101 and A.103 we can extract 4 equations, where there are 4 partial derivatives that are known ($\frac{\partial u_u^*}{\partial \kappa_1}$, $\frac{\partial v_u^*}{\partial \kappa_1}$, $\frac{\partial u_u^*}{\partial \kappa_2}$ and $\frac{\partial v_u^*}{\partial \kappa_2}$); and 4 unknown partial derivatives that are $\frac{\partial u_d^*}{\partial \kappa_1}$, $\frac{\partial v_d^*}{\partial \kappa_1}$, $\frac{\partial u_d^*}{\partial \kappa_2}$ and $\frac{\partial v_d^*}{\partial \kappa_2}$. Solving for them we have $\frac{\partial \mathbf{h}_d^*}{\partial \kappa_1}$ and $\frac{\partial \mathbf{h}_d^*}{\partial \kappa_2}$, that we can use to compute the entire derivative of the image measurements with respect to the radial distortion

$$\frac{\partial \mathbf{h}_d}{\partial \kappa_1} = \frac{\partial \mathbf{h}_d}{\partial \mathbf{h}_d^*}\frac{\partial \mathbf{h}_d^*}{\partial \kappa_1} \tag{A.105}$$

$$\frac{\partial \mathbf{h}_d}{\partial \kappa_2} = \frac{\partial \mathbf{h}_d}{\partial \mathbf{h}_d^*}\frac{\partial \mathbf{h}_d^*}{\partial \kappa_2}. \tag{A.106}$$

A.3.3 Inverse Depth Point Feature Initialization and Derivatives

As it happened in the previous section, point initialization function is exactly the same as in the calibrated case (equations A.56 to A.62), except for the introduction of the intermediate variables $\mathbf{h}_u^*$ and $\mathbf{h}_d^*$ of the previous section. The image point to be initialized $\mathbf{h}_d$ is first then converted to the intermediate variable $\mathbf{h}_d^*$

$$\mathbf{h}_d^* = \mathbf{h}_d - \begin{pmatrix} C_x \\ C_y \end{pmatrix}. \tag{A.107}$$

After that, the undistortion model for the lens is applied

$$\mathbf{h}_u^* = \begin{pmatrix} u_u^* \\ v_u^* \end{pmatrix} = \mathbf{h}_d^*\left(1+\kappa_1 r_d^2+\kappa_2 r_d^4\right) \tag{A.108}$$

$$r_d = \sqrt{\left(d_x u_d^*\right)^2 + \left(d_y v_d^*\right)^2}\,; \tag{A.109}$$

And finally the image point is backprojected into a 3D point in the camera reference frame at unit depth in the optical axis

$$\mathbf{h}^C = \begin{pmatrix} \frac{-u_u^* d_x}{f} \\ \frac{-v_u^* d_y}{f} \\ 1 \end{pmatrix} . \qquad \text{(A.110)}$$

From here, the rest of the steps to obtain the newly initialized inverse depth feature $\mathbf{y}^{NEW}$ are the same than those in section A.2.3 from equation A.59 to A.62.

The derivatives now accounts for the fact that calibration is included now in the state vector. The Jacobian of the initialization function is in this case

$$\mathbf{J} = \begin{pmatrix} & \mathbf{I} & \begin{matrix} 0 \\ \vdots \\ 0 \end{matrix} \\ \frac{\partial \mathbf{y}}{\partial \mathbf{x}_C} & 0 \ldots 0 & \frac{\partial \mathbf{y}}{\partial \mathbf{h}} \end{pmatrix} , \qquad \text{(A.111)}$$

where the derivatives by the camera $\frac{\partial \mathbf{y}}{\partial \mathbf{x}_C}$ now include the internal calibration variables

$$\frac{\partial \mathbf{y}}{\partial \mathbf{x}_C} = \begin{pmatrix} \frac{\partial \mathbf{y}}{\partial f} & \frac{\partial \mathbf{y}}{\partial C_x} & \frac{\partial \mathbf{y}}{\partial C_y} & \frac{\partial \mathbf{y}}{\partial \kappa_1} & \frac{\partial \mathbf{y}}{\partial \kappa_2} & \frac{\partial \mathbf{y}}{\partial \mathbf{r}^{WC}} & \frac{\partial \mathbf{y}}{\partial \mathbf{q}^{WC}} & \cancelto{0}{\frac{\partial \mathbf{y}}{\partial v}} & \cancelto{0}{\frac{\partial \mathbf{y}}{\partial \omega}} \end{pmatrix} . \qquad \text{(A.112)}$$

From the six parameters defining an inverse depth feature, only the azimuth-elevation angles defining the ray are the ones depending on the calibration in the initialization step. The derivative by the calibration is then

$$\frac{\partial \mathbf{y}}{\partial f} = \begin{pmatrix} 0 & 0 & 0 & \frac{\partial \theta}{\partial f} & \frac{\partial \phi}{\partial f} & 0 \end{pmatrix}^\top \qquad \text{(A.113)}$$

$$\frac{\partial \mathbf{y}}{\partial C_x} = \begin{pmatrix} 0 & 0 & 0 & \frac{\partial \theta}{\partial C_x} & \frac{\partial \phi}{\partial C_x} & 0 \end{pmatrix}^\top \qquad \text{(A.114)}$$

$$\frac{\partial \mathbf{y}}{\partial C_y} = \begin{pmatrix} 0 & 0 & 0 & \frac{\partial \theta}{\partial C_y} & \frac{\partial \phi}{\partial C_y} & 0 \end{pmatrix}^\top \qquad \text{(A.115)}$$

$$\frac{\partial \mathbf{y}}{\partial \kappa_1} = \begin{pmatrix} 0 & 0 & 0 & \frac{\partial \theta}{\partial \kappa_1} & \frac{\partial \phi}{\partial \kappa_1} & 0 \end{pmatrix}^\top \qquad \text{(A.116)}$$

$$\frac{\partial \mathbf{y}}{\partial \kappa_2} = \begin{pmatrix} 0 & 0 & 0 & \frac{\partial \theta}{\partial \kappa_2} & \frac{\partial \phi}{\partial \kappa_2} & 0 \end{pmatrix}^\top . \qquad \text{(A.117)}$$

The above derivatives of azimuth-elevation angle pair by calibration can be computed via the chain rule. The derivatives by the focal length are as follows,

$$\frac{\partial \theta}{\partial f} = \frac{\partial \theta}{\partial \mathbf{h}^W} \frac{\partial \mathbf{h}^W}{\partial \mathbf{h}^C} \frac{\partial \mathbf{h}^C}{\partial f} \tag{A.118}$$

$$\frac{\partial \theta}{\partial \mathbf{h}^W} = \begin{pmatrix} \frac{h_z^W}{h_x^{W2} + h_z^{W2}} & 0 & \frac{-h_x^W}{h_x^{W2} + h_z^{W2}} \end{pmatrix} \tag{A.119}$$

$$\frac{\partial \mathbf{h}^W}{\partial \mathbf{h}^C} = \mathbf{R}^{WC} \tag{A.120}$$

$$\frac{\partial \mathbf{h}^C}{\partial f} = \begin{pmatrix} \frac{-(u_u - C_x)d_x}{f^2} \\ \frac{-(v_u - C_y)d_y}{f^2} \\ 0 \end{pmatrix} \tag{A.121}$$

$$\frac{\partial \phi}{\partial f} = \frac{\partial \phi}{\partial \mathbf{h}^W} \frac{\partial \mathbf{h}^W}{\partial \mathbf{h}^C} \frac{\partial \mathbf{h}^C}{\partial f} \tag{A.122}$$

$$\frac{\partial \phi}{\partial \mathbf{h}^W} = \begin{pmatrix} \frac{h_x^W h_y^W}{\left(h_x^{W2} + h_y^{W2} + h_z^{W2}\right)\sqrt{h_x^{W2} + h_z^{W2}}} \\ \frac{-\sqrt{h_x^{W2} + h_z^{W2}}}{\left(h_x^{W2} + h_y^{W2} + h_z^{W2}\right)} \\ \frac{h_y^W h_z^W}{\left(h_x^{W2} + h_y^{W2} + h_z^{W2}\right)\sqrt{h_x^{W2} + h_z^{W2}}} \end{pmatrix}^\top . \tag{A.123}$$

The derivatives by the radial distortion parameters are easily extracted

$$\frac{\partial \theta}{\partial \kappa_1} = \frac{\partial \theta}{\partial \mathbf{h}^W} \frac{\partial \mathbf{h}^W}{\partial \mathbf{h}^C} \frac{\partial \mathbf{h}^C}{\partial \kappa_1} \tag{A.124}$$

$$\frac{\partial \mathbf{h}^C}{\partial \kappa_1} = \begin{pmatrix} \frac{-d_x}{f}(u_d - C_x) r_d^2 & \frac{-d_y}{f}(v_d - C_y) r_d^2 & 0 \end{pmatrix}^\top \tag{A.125}$$

$$\frac{\partial \phi}{\partial \kappa_1} = \frac{\partial \phi}{\partial \mathbf{h}^W} \frac{\partial \mathbf{h}^W}{\partial \mathbf{h}^C} \frac{\partial \mathbf{h}^C}{\partial \kappa_1} \tag{A.126}$$

$$\frac{\partial \theta}{\partial \kappa_2} = \frac{\partial \theta}{\partial \mathbf{h}^W} \frac{\partial \mathbf{h}^W}{\partial \mathbf{h}^C} \frac{\partial \mathbf{h}^C}{\partial \kappa_2} \tag{A.127}$$

$$\frac{\partial \mathbf{h}^C}{\partial \kappa_2} = \begin{pmatrix} \frac{-d_x}{f}(u_d - C_x) r_d^4 & \frac{-d_y}{f}(v_d - C_y) r_d^4 & 0 \end{pmatrix}^\top \tag{A.128}$$

$$\frac{\partial \phi}{\partial \kappa_2} = \frac{\partial \phi}{\partial \mathbf{h}^W} \frac{\partial \mathbf{h}^W}{\partial \mathbf{h}^C} \frac{\partial \mathbf{h}^C}{\partial \kappa_2} . \tag{A.129}$$

And finally, the derivatives by the principal point

$$\mathbf{h}_u^* = \begin{pmatrix} u_u^* \\ v_u^* \end{pmatrix} = \begin{pmatrix} u_u - C_x \\ v_u - C_y \end{pmatrix} \tag{A.130}$$

$$\frac{\partial \theta}{\partial C_x} = \frac{\partial \theta}{\partial \mathbf{h}^W} \frac{\partial \mathbf{h}^W}{\partial \mathbf{h}^C} \frac{\partial \mathbf{h}^C}{\partial \mathbf{h}_u^*} \frac{\partial \mathbf{h}_u^*}{\partial C_x} \tag{A.131}$$

$$\frac{\partial \mathbf{h}^C}{\partial \mathbf{h}_u^*} = \begin{pmatrix} \frac{-d_x}{f} & 0 \\ 0 & \frac{-d_y}{f} \\ 0 & 0 \end{pmatrix} \tag{A.132}$$

$$\frac{\partial \mathbf{h}_u^*}{\partial C_x} = \begin{pmatrix} \frac{\partial u_u^*}{\partial C_x} \\ \frac{\partial v_u^*}{\partial C_x} \end{pmatrix} \tag{A.133}$$

$$\frac{\partial u_u^*}{\partial C_x} = -\left(1 + \kappa_1 r_d^2 + \kappa_2 r_d^4 + (u_d - C_x)^2 d_x^2 \left(\kappa_1 + 2\kappa_2 r_d^2\right)\right) \tag{A.134}$$

$$\frac{\partial v_u^*}{\partial C_x} = -\left((v_d - C_y)(u_d - C_x) d_x^2 \left(\kappa_1 + 2\kappa_2 r_d^2\right)\right) \tag{A.135}$$

$$\frac{\partial \phi}{\partial C_x} = \frac{\partial \phi}{\partial \mathbf{h}^W} \frac{\partial \mathbf{h}^W}{\partial \mathbf{h}^C} \frac{\partial \mathbf{h}^C}{\partial \mathbf{h}_u^*} \frac{\partial \mathbf{h}_u^*}{\partial C_x} \tag{A.136}$$

$$\frac{\partial \theta}{\partial C_y} = \frac{\partial \theta}{\partial \mathbf{h}^W} \frac{\partial \mathbf{h}^W}{\partial \mathbf{h}^C} \frac{\partial \mathbf{h}^C}{\partial \mathbf{h}_u^*} \frac{\partial \mathbf{h}_u^*}{\partial C_y} \tag{A.137}$$

$$\frac{\partial \mathbf{h}_u^*}{\partial C_y} = \begin{pmatrix} \frac{\partial u_u^*}{\partial C_y} \\ \frac{\partial v_u^*}{\partial C_y} \end{pmatrix} \tag{A.138}$$

$$\frac{\partial u_u^*}{\partial C_y} = -\left((u_d - C_x)(v_d - C_y) d_y^2 \left(\kappa_1 + 2\kappa_2 r_d^2\right)\right) \tag{A.139}$$

$$\frac{\partial v_u^*}{\partial C_y} = -\left(1 + \kappa_1 r_d^2 + \kappa_2 r_d^4 + (v_d - C_y)^2 d_y^2 \left(\kappa_1 + 2\kappa_2 r_d^2\right)\right) \tag{A.140}$$

A.4 Quaternion Normalization

Quaternion normalization must be performed after each update step of the Extended Kalman Filter in order to ensure that its norm equals one, i.e. $q_0^2 + q_1^2 + q_2^2 + q_3^2 = 1$. The state vector is then modified as follows

$$\mathbf{x}^{norm} = \begin{pmatrix} \mathbf{r}^{WC} \\ \frac{\mathbf{q}^{WC}}{|\mathbf{q}^{WC}|} \\ \mathbf{v}^W \\ \omega^C \\ \mathbf{x}_{map} \end{pmatrix}, \tag{A.141}$$

and the covariance should be updated with the Jacobian of the transformation

$$\mathbf{P}^{norm} = \mathbf{J}_{norm}\mathbf{P}\mathbf{J}_{norm}^{\top}, \tag{A.142}$$

$$\mathbf{J}_{norm} = \begin{pmatrix} I & \mathbf{0} & \mathbf{0} \\ \mathbf{0} & \frac{\partial \mathbf{q}^{norm}}{\partial \mathbf{q}} & \mathbf{0} \\ \mathbf{0} & \mathbf{0} & I \end{pmatrix}, \tag{A.143}$$

$$\frac{\partial \mathbf{q}^{norm}}{\partial \mathbf{q}} = \left(q_0^2+q_1^2+q_2^2+q_3^2\right)^{-\frac{3}{2}} \times \tag{A.144}$$

$$\times \begin{pmatrix} q_1^2+q_2^2+q_3^2 & -q_0q_1 & -q_0q_2 & -q_0q_3 \\ -q_1q_0 & q_0^2+q_2^2+q_3^2 & -q_1q_2 & -q_1q_3 \\ -q_2q_0 & -q_2q_1 & q_0^2+q_1^2+q_3^2 & -q_2q_3 \\ -q_3q_0 & -q_3q_1 & -q_3q_2 & q_0^2+q_1^2+q_2^2 \end{pmatrix} \tag{A.145}$$

A.5 Inverse Depth to Cartesian Parameterization Conversion

As detailed in section 3.2.3, each inverse depth feature $y_{\rho,i}$ represents a Euclidean point $(X_i, Y_i, Z_i)^{\top}$ that can be computed as follows

$$\mathbf{y}_{XYZ,i} = \begin{pmatrix} X_i \\ Y_i \\ Z_i \end{pmatrix} = \mathbf{g}_i\left(\mathbf{y}_{\rho,i}\right) = \begin{pmatrix} x_i \\ y_i \\ z_i \end{pmatrix} + \frac{1}{\rho_i}\mathbf{m}_i\left(\theta_i, \phi_i\right), \tag{A.146}$$

being $\mathbf{m}_i = \left(\cos\phi_i \sin\theta_i \quad -\sin\phi_i \quad \cos\phi_i \cos\theta_i\right)^{\top}$.

In the algorithms described in the book, when an inverse depth feature $y_{\rho,i}$ holds the conversion criteria detailed in section 3.6, it can be safely converted to a Cartesian –XYZ– feature $y_{XYZ,i}$ without degrading the EKF performance. The state vector $\mathbf{x} = \left(\mathbf{x}_C, \mathbf{y}_1^{\top}, \ldots, \mathbf{y}_{\rho,i}^{\top}, \ldots, \mathbf{y}_n^{\top}\right)^{\top}$ is then transformed into $\mathbf{x}^{NEW} = \left(\mathbf{x}_C, \mathbf{y}_1^{\top}, \ldots, \mathbf{y}_{XYZ,i}^{\top}, \ldots, \mathbf{y}_n^{\top}\right)^{\top}$, where $\mathbf{y}_{XYZ,i}^{\top}$ comes from the conversion described before in equation A.146.

The state vector covariance is transformed by using the Jacobian of the transformation $\mathbf{J}_{g_i}$, being the new covariance $\mathbf{P}^{NEW} = \mathbf{J}_{g_i}\mathbf{P}\mathbf{J}_{g_i}^{\top}$. The relevant terms of the Jacobian $\mathbf{J}_{g_i}$ are in the position of feature i

$$\mathbf{J}_{g_i} = \begin{pmatrix} I & \mathbf{0} & \mathbf{0} \\ \mathbf{0} & \frac{\partial \mathbf{y}_{XYZ,i}}{\partial \mathbf{y}_{\rho,i}} & \mathbf{0} \\ \mathbf{0} & \mathbf{0} & I \end{pmatrix}. \tag{A.147}$$

The derivative $\frac{\partial \mathbf{y}_{XYZ,i}}{\partial \mathbf{y}_{\rho,i}}$ can be divided into the following parts

$$\frac{\partial \mathbf{y}_{XYZ,i}}{\partial \mathbf{y}_{\rho,i}} = \left(\frac{\partial \mathbf{y}_{XYZ,i}}{\partial \mathbf{r}_i} \frac{\partial \mathbf{y}_{XYZ,i}}{\partial \theta_i} \frac{\partial \mathbf{y}_{XYZ,i}}{\partial \phi_i} \frac{\partial \mathbf{y}_{XYZ,i}}{\partial \rho_i} \right) ; \tag{A.148}$$

where each of those is as follows

$$\frac{\partial \mathbf{y}_{XYZ,i}}{\partial \mathbf{r}_i} = \mathbf{I} \tag{A.149}$$

$$\frac{\partial \mathbf{y}_{XYZ,i}}{\partial \theta_i} = \frac{\partial \mathbf{y}_{XYZ,i}}{\partial \mathbf{m}} \frac{\partial \mathbf{m}}{\partial \theta_i} \tag{A.150}$$

$$\frac{\partial \mathbf{y}_{XYZ,i}}{\partial \phi_i} = \frac{\partial \mathbf{y}_{XYZ,i}}{\partial \mathbf{m}} \frac{\partial \mathbf{m}}{\partial \phi_i} \tag{A.151}$$

$$\frac{\partial \mathbf{y}_{XYZ,i}}{\partial \rho_i} = \frac{-\mathbf{m}}{\rho^2} . \tag{A.152}$$

Here $\frac{\partial \mathbf{y}_{XYZ,i}}{\partial \mathbf{m}} = \frac{1}{\rho}\mathbf{I}$, and $\frac{\partial \mathbf{m}}{\partial \theta_i}$ and $\frac{\partial \mathbf{m}}{\partial \phi_i}$ were already detailed in A.54 and A.54.

Appendix B
Filter Tuning Understanding via Dimensional Analysis

B.1 Introduction

It is a well known fact in SfM that a moving calibrated camera observing a scene can recover scene geometry and camera motion only up to a scale factor — scene scale is an non-observable magnitude if only bearing measurements are made. Unlike SfM, previous SLAM work [Davison 2003, Montiel *et al.* 2006] have used extra information in the form of a known initialisation object to fix the scene scale.

In this appendix it is given insight on how this non-visual information is in fact not essential for solving the tracking problem and that no known target object needs to be added to the scene. While this means that an overall scene scale cannot be recovered, real-time tracking can still proceed. And if at any time extra information concerning the real scale does become available, it can be incorporated into the estimation.

In particular, a novel understanding of the EKF-based SfM problem in terms of dimensionless parameters is derived using Buckingham's Π theorem [Buckingham 1914]. Π theorem relies on the requirement for dimensional correctness in any formula and hence also any estimation process. Our EKF SfM algorithm therefore recovers dimensionless, up-to-scale geometry, and also provides benefits by allowing previous tuning parameters to be rolled up into a canonical set which give an important new understanding of the uncertainties in the system now in pixel units. These parameters in the image provide a natural way of understanding image sequences, irrespectively of the frame rate, actual scene and camera motion.

Further, it is also shown that jointly with the main dimensionless part of the SLAM state vector an extra parameter representing metric scale could be added. During tracking, vision-only measurements would not reduce the uncertainty in the scale parameter but only in the dimensionless scene geometry. However, any measurement containing metric information such as odometry, a feature at a known depth or the distance between two features can be added when available and will correct both the scale and the dimensionless scene geometry.

B.2 Monocular SLAM Estimation Process

As in the previous chapters of the book, the state vector is divided into the camera parameters $\mathbf{x}_C$ and the parameters corresponding to the n features in the map $\mathbf{y}_i$.

$$\mathbf{x} = \left(\mathbf{x}_C^\top \; \mathbf{y}_1^\top \; \dots \; \mathbf{y}_i^\top \; \dots \; \mathbf{y}_n^\top \right)^\top . \tag{B.1}$$

The camera state vector includes camera position, quaternion orientation, and linear and angular velocities:

$$\mathbf{x}_C = \left(\mathbf{r}^{WC^\top} \; \mathbf{q}^{WC^\top} \; \mathbf{v}^{W^\top} \; \omega^{C^\top} \right)^\top . \tag{B.2}$$

Points are coded in inverse depth parameterization; as proposed in chapter 3 (see section 3.2.3 for details)

$$\mathbf{y_i} = \left(\mathbf{r}_i^\top \; \theta_i \; \phi_i \; \rho_i \right)^\top = \left(x_i \; y_i \; z_i \; \theta_i \; \phi_i \; \rho_i \right)^\top . \tag{B.3}$$

From here, we will split the state vector into a metric parameter d –unobservable from image measurements– and a dimensionless state representing the scene and camera parameters. Doing this, the state vector is partitioned according to observability with a monocular camera. Camera measurements will reduce the scene geometric uncertainty on the dimensionless parameters; but not the uncertainty in the metric parameter d.

$$\mathbf{x} = \left(d \;\; \Pi_r^\top \;\; \mathbf{q}^{WC^\top} \;\; \Pi_v^\top \;\; \Pi_\omega^\top \;\; \Pi_{y_1}^\top \;\; \dots \;\; \Pi_{y_i}^\top \;\; \dots \;\; \Pi_{y_n}^\top \right)^\top . \tag{B.4}$$

Notice that the orientation parameters in $\mathbf{q}^{WC}$ are already dimensionless, and hence remains the same in this dimensionless formulation.

The mapping from this dimensionless state vector to a metric one involves a non-linear computation using the dimensionless geometry and the metric parameter d:

$$\mathbf{r} = d\,\Pi_r \tag{B.5}$$

$$\mathbf{v} = d\,\Pi_v \tag{B.6}$$

$$\omega = \Pi_\omega \tag{B.7}$$

$$\mathbf{y}_i = \begin{pmatrix} d\,\Pi_{xi} \\ d\,\Pi_{yi} \\ d\,\Pi_{zi} \\ \theta_i \\ \phi_i \\ \Pi_{\rho i}/d \end{pmatrix} . \tag{B.8}$$

B.3 Buckingham's Π Theorem Applied to Monocular SLAM

Buckingham's Π Theorem [Buckingham 1914] is a key theorem in Dimensional Analysis. It basically states that the physical laws hold independently of the specific system of units that is chosen. Hence, given a dimensionally correct equation involving n quantities of different kinds $f(X_1, X_2, X_3, ..., X_n) = 0$, the existing relationship between the variables can also be expressed as $F(\Pi_1, \Pi_2, \Pi_3, ..., \Pi_{n-k}) = 0$ where Π_i is a reduced set of $n-k$ independent dimensionless groups of variables, and k the number of independent dimensions appearing in the problem.

The monocular estimation process can be expressed as the following function:

$$\left(\mathbf{r}^\top \; \mathbf{q}^\top \; \mathbf{v}^\top \; \omega^\top \; \mathbf{y}_1^\top \; \ldots \; \mathbf{y}_n^\top\right)^\top = \mathbf{f}(\sigma_a, \sigma_\alpha, \mathbf{z}, \sigma_z, \Delta t, \rho_0, \sigma_{\rho_0}, \sigma_{v0}, \sigma_{\omega 0}) \,, \tag{B.9}$$

where vector $\mathbf{z}$ stacks all the image measurements along the image sequence. Table B.1 summarizes all the variables involved in monocular SLAM estimation and and their units.

Table B.1. Variables involved in the monocular SLAM estimation and their dimensions

$\mathbf{r}, \mathbf{r}_i$	$\mathbf{q}$	$\mathbf{v}$, σ_{v0}	ω, $\sigma_{\omega 0}$	σ_a	σ_α	θ_i, ϕ_i	ρ_i, ρ_0, σ_{ρ_0}	$\mathbf{z}, \sigma_z$	Δt
l	1	lt^{-1}	t^{-1}	lt^{-2}	t^{-2}	1	l^{-1}	1	t

Based on table B.1, the dimensionless groups must be chosen. It can be seen that the dimensions involved in the SLAM estimation are space and time, so two variables containing these dimensions should be chosen to form the dimensionless groups. We have chosen the parameters ρ_0 and Δt; and the corresponding dimensionless variables can be seen in (Table B.2).

Table B.2. Dimensionless numbers and the corresponding involved variables

Π_r	Π_{r_i}	Π_q	Π_v	$\Pi_{\sigma_{v0}}$	Π_ω	$\Pi_{\sigma_{\omega 0}}$	$\Pi_{\sigma a}$	$\Pi_{\sigma\alpha}$	Π_{ρ_i}	$\Pi_{\sigma_{\rho 0}}$	Π_z	Π_{σ_z}
$\mathbf{r}\rho_0$	$\mathbf{r}_i\rho_0$	$\mathbf{q}$	$\mathbf{v}\rho_0\Delta t$	$\sigma_{v0}\rho_0\Delta t$	$\omega\Delta t$	$\sigma_{\omega 0}\Delta t$	$\sigma_a\rho_0\Delta t^2$	$\sigma_\alpha\Delta t^2$	$\frac{\rho_i}{\rho_0}$	$\frac{\sigma_{\rho 0}}{\rho_0}$	$\mathbf{z}$	σ_z

B.4 Dimensionless Monocular SLAM Model

Using the dimensionless magnitudes, our new monocular SLAM model can be defined as follows. The state vector is composed of the dimensionless parameters defining the camera location, rotation and velocities; and the parameters related with the map features:

$$\Pi_{x_C} = \left(\Pi_r{}^\top, \mathbf{q}^\top, \Pi_v{}^\top, \Pi_\omega{}^\top\right)^\top \quad \Pi_{y_i} = \left(\Pi_{r_i}{}^\top, \theta_i, \phi_i, \Pi_{\rho_i}\right)^\top \tag{B.10}$$

The dynamic model for the camera motion in this dimensionless formulation is The dimensionless state update equation is:

$$\mathbf{f}_v = \begin{pmatrix} \Pi_{rk+1} \\ \mathbf{q}_{k+1} \\ \Pi_{vk+1} \\ \Pi_{\omega k+1} \end{pmatrix} = \begin{pmatrix} \Pi_{rk} + \Pi_{vk} + \Pi_{ak} \\ \mathbf{q}_k \times \mathbf{q}(\Pi_{\omega k} + \Pi_{\alpha k}) \\ \Pi_{vk} + \Pi_{ak} \\ \Pi_{\omega k} + \Pi_{\alpha k} \end{pmatrix} . \tag{B.11}$$

The insight about the geometrical meaning of the parameters in this equation will be given in next sections. In the measurement model, the features coded in inverse depth must be converted to 3D points in the world reference first:

$$\Pi_h^W = \Pi_{r_i} + \frac{1}{\Pi_{\rho_i}} \mathbf{m}(\theta_i, \phi_i) , \tag{B.12}$$

where $\mathbf{m}(\theta_i, \phi_i)$ is the unit vector defined by the azimuth-elevation pair. The 3D points in the world reference are then converted to the camera frame:

$$\Pi_h^C = \mathbf{R}^{CW}(\mathbf{q})\,(\Pi_h^W - \Pi_r) \tag{B.13}$$

and projected into the camera using the pinhole model:

$$\upsilon = \frac{\Pi_h^C|_x}{\Pi_h^C|_z} \quad \nu = \frac{\Pi_h^C|_y}{\Pi_h^C|_z} \tag{B.14}$$

It is worth remarking that the measurement equation do not involve the actual size of the scene, coded in the metric parameter d. Visual measurements from a monocular camera would not reduce then the initial uncertainty in the size of the scene. If the metric size had to be estimated, other measurements should be made. For example, if a distance between two points is known, it can be used for reducing the uncertainty over d:

$$\mathscr{D}(\mathbf{P_1}, \mathbf{P_2}) = d\sqrt{(\Pi_{y_2}|_x - \Pi_{y_1}|_x)^2 + (\Pi_{y_2}|_y - \Pi_{y_1}|_y)^2 + (\Pi_{y_2}|_z - \Pi_{y_1}|_z)^2} . \tag{B.15}$$

B.5 Geometric Interpretation of the Dimensionless Parameters

Figure B.1 details the geometrical interpretation of the main dimensionless parameters that are involved in the estimation. The dimensionless parameters obtained in sections B.3 and B.4 actually represent image quantities in pixel units, allowing an understanding of the filter tuning parameters as image motion quantities instead of parameters in the 3D world.

As the input to a monocular SLAM system is an image sequence, the motion between two images can be easily extracted from the sequence but it may be hard to guess in terms of 3D quantities. And in addition to that, it is the ratios of the 3D parameters the numbers that are important for the estimation. For example, the image motion is the same for a certain scene and camera translation than for a scene with double size and a camera that also translates double. The image motion also would be the same if the frame rate is double but the camera moves twice faster. Each one of these four configurations should use different filter tuning parameters in a standard formulation of monocular SLAM; but they are nicely represented by a unique tuning in our dimensionless formulation as they imply the same image motion.

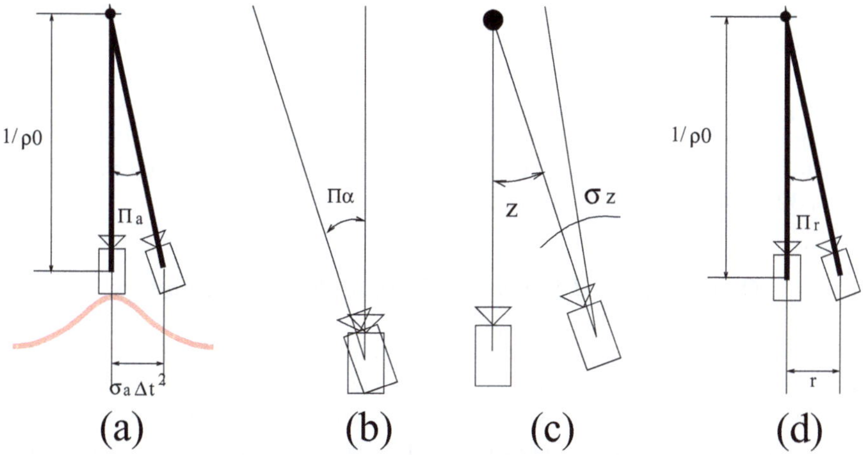

Fig. B.1. Geometric interpretation of the dimensionless monocular SLAM parameters.

Figure B.1(a) illustrates the geometric meaning ot the dimensionless parameter Π_{σ_a}. The product $\sigma_a \Delta t^2$ represents the effect of the acceleration noise on the camera location. This value divided by $1/\rho_0$ gives the angle represented in the figure. This angle can be seen as the parallax allowed to a feature at depth $1/\rho_0$ due to camera acceleration.

The camera angular acceleration covariance in Figure B.1(b) can clearly be interpreted as an angle between frames and can be measured –as calibration is known– in image pixels. Image measurements and image noise, in Figure B.1(c) are directly measured in the image, so they are already equivalent to dimensionless angles.

The translation estimate, Π_r (in fig B.1(d)), can also be seen as the angle defined by the translation between frames and the initial inverse depth.

As a result of the analysis given in this section, and with minimum changes in the structure of the EKF estimation scheme, all estimated parameters and filter tuning can be seen as dimensionless angles and consequently transformed to pixels in a calibrated camera. We believe that this approach simplifies the filter tuning: 3D acceleration values, that cannot be directly extracted from the visualization of the sequence to be processed are no longer required. Instead of that, quantities measured in pixels in the image are the only numbers necessary for the tuning –easier to extract from the sequence.

B.6 Real Image Results

We have performed several real-image experiments in order to show the main advantages of this new interpretation of the monocular SLAM problem under a dimensionless optic. The first one illustrates how the scale of the scene in usual monocular SLAM depends on the prior knowledge added to scene, even if this knowledge is very weak –in the case where no known target is added, scale depends on prior inverse depth values and acceleration noise. The second experiment shows the use of image tuning and the reduction in the number of tuning parameters that comes as consequence from the proposed scheme. In the third experiment, the same image tuning is used in two different sequences which have different metric qualities but lead to the same image motion. All of the sequences have been recorded with a IEEE 1394 320×240 monochrome camera at 30 fps.

B.6.1 Dependence of Scene Scale on a Priori Parameters

The same sequence was processed with the dimensional EKF SLAM algorithm varying the ρ_0 parameter. Figure B.2 shows the estimation for $\rho_0 = 0.5m^{-1}$ and $\rho_0 = 0.1m^{-1}$. Notice that the estimated depth of the scene (the distance between the camera and the points in the bookcase) tends to be at the depth prior (2m and 10m). The two estimated scenes have the same form, the difference is just the scale of the

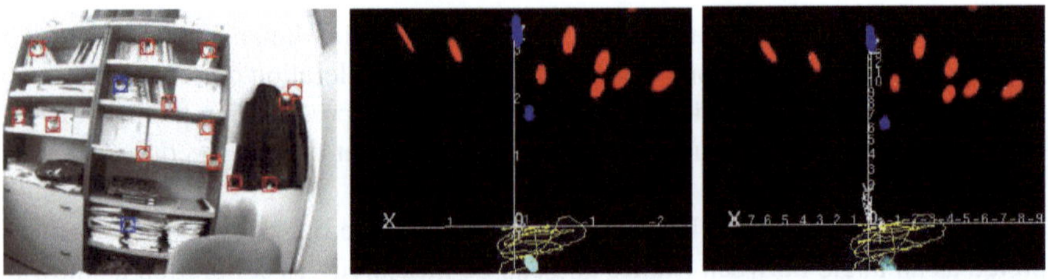

Fig. B.2. Left: sample. Centre: EKF SLAM estimation result $\rho_0 = 0.5m^{-1}$. Right: EKF SLAM estimation result $\rho_0 = 0.1m^{-1}$. The uncertainty for the features is plotted in red and blue, the camera uncertainty in cyan and the camera trajectory in yellow.

axis. If $\Pi_r^{WC} = \rho_0 \mathbf{r}^{\mathbf{WC}}$ and Π_{y_i} were estimated using the dimensionless monocular SLAM proposed, these two experiments would be normalized into one, in which normalized depth tends to be at *dimensionless* '1',

B.6.2 Image Tuning in a Pure Rotation Sequence

This sequence is a pure camera rotation in a hallway. Dimensional monocular SLAM should have been tuned with real camera accelerations and depth priors. As these values are not observable, they need to be guessed. In the dimensionless monocular SLAM of this chapter, the filter parameters are tuned in image units, which are directly observable.

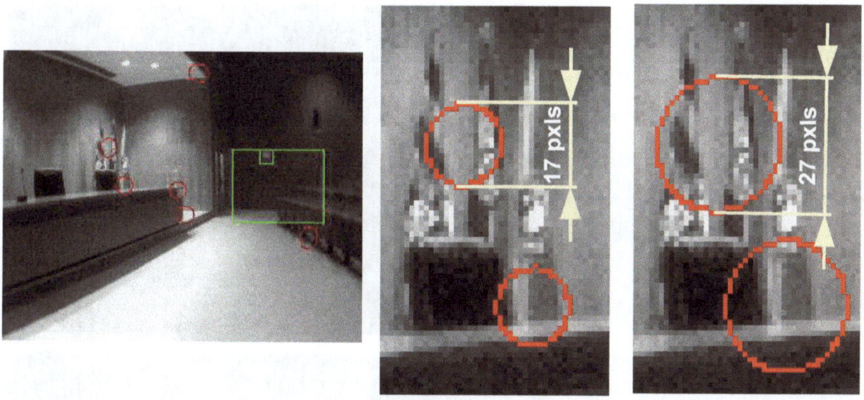

Fig. B.3. Pure rotation image search regions. Left: sample image. Centre: $\Pi_\alpha = 2 pixels$. Right:$\Pi_\alpha = 4 pixels$.

Two experiments with the same $\Pi_{\sigma a} = 0$, $\Pi_{\sigma z} = 1 pixels$ values but different tuning in $\Pi_{\sigma\alpha}$: a) $\Pi_{\sigma\alpha} = 2 pixels$, and b) $\Pi_{\sigma\alpha} = 4 pixels$ has been performed (figure B.3). Because of the image tuning, their effect can be directly seen in the 95% image search regions size for the map features.

It is important to notice again here that neither 3D scene assumptions nor time between frames Δt have been neded in the previous paragraph to propose a filter tuning. The tuned values are the allowed image motion between frames due to camera linear and angular acceleration and image noise.

B.6.3 The Same Image Tuning for Different Sequences

Two translational sequences have been recorded walking along a corridor and looking at the wall. In the first one, the distance from the wall was 2.5 metres. In the second, the distance from the wall was twice (5 metres), the distance walked along the corridor the same, and the walking velocity was double (therefore, the number of frames of the second sequence is half the first one). Although they are two different experiments, the image motion in both sequences is the same, and dimensionless monocular SLAM has to be tuned with same values. In this experiment, these values were: $\sigma_z = 1pxl$, $\sigma_a = 2pxl$ and $\sigma_\alpha = 2pxl$. Notice again the simplicity of image tuning compared with 3D tuning, in which you have to guess the unobservable depth prior and the 3D accelerations. Figure B.4 shows the results of both estimations.

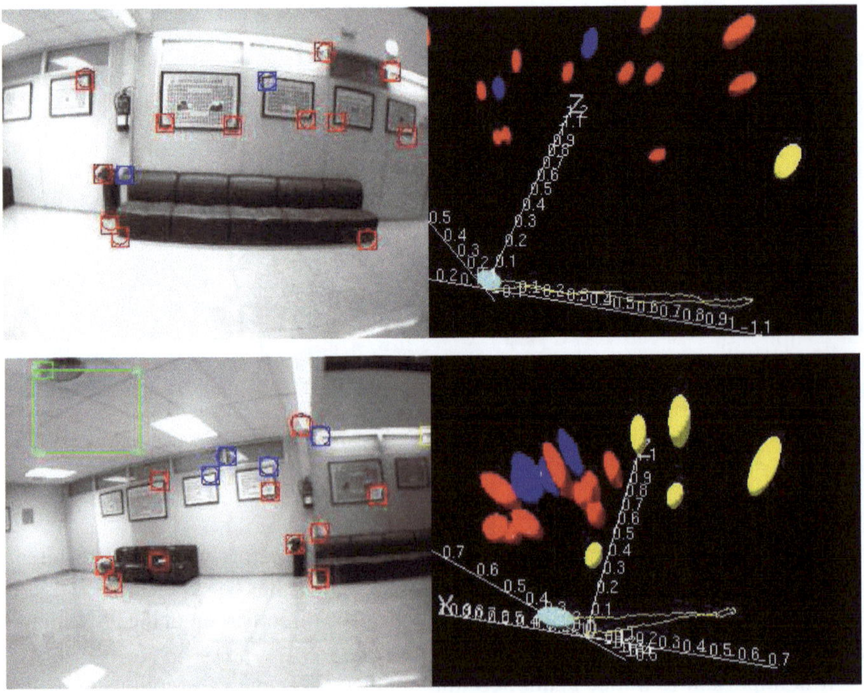

Fig. B.4. Two equivalent sequences. First and last images and 3D estimated geometry.

The dimensionless estimated translation can be interpreted as the translation in units of the initial depth prior. As the wall is twice as far in the second sequence, the second sequence's estimated translation is half. It can also be noticed that, as the normalized translation is smaller in the second experiment, the normalized 3D point positions are estimated with less accuracy and have larger uncertainty regions.

B.7 Discussion

As a result of an analysis of the SfM problem under Π Buckingham's theorem, a dimensionless parametrization for the EKF state vector can be proposed that clarifies the role of the filter tuning parameters as image motion quantities. As a consequence, filter tuning can be done using the allowed accelerations to the filter projected in the images –that is, in pixels– instead of using the allowed acceleration itself in metric units –which is more difficult to estimate having the sequence as the only input to the algorithm.

Up-to-scale results from real-time, EKF based monocular SLAM without an initialisation target are presented. As no known points are included in the estimation, the real size of the scene is not going to be recovered. A scaled estimation is going to be obtained, being the overall scene scale being determined from a priori knowledge inserted to the filter –even if it is so weak as inverse depth priors and linear acceleration noise are.

References

[2d3 2011] 2d3 (June 2011), `http://www.2d3.com/`

[Aidala & Hammel 1983] Aidala, V.J., Hammel, S.E.: Utilization of Modified Polar Coordinates for Bearing-Only Tracking. IEEE Transactions on Automatic Control 28(3), 283–294 (1983)

[Alspach & Sorenson 1972] Alspach, D., Sorenson, H.: Nonlinear Bayesian estimation using Gaussian sum approximations. IEEE Transactions on Automatic Control 17(4), 439–448 (1972)

[Autosticht 2011] Autosticht (June 2011), `http://www.autostitch.net/`

[Ayache & Faugeras 1989] Ayache, N., Faugeras, O.D.: Maintaining representations of the environment of a mobile robot. IEEE Transactions on Robotics and Automation 5(6), 804–819 (1989)

[Azarbayejani & Pentland 1995] Azarbayejani, A., Pentland, A.P.: Recursive estimation of motion, structure, and focal length. IEEE Transactions on Pattern Analysis and Machine Intelligence 17(6), 562–575 (1995)

[Bailey & Durrant-Whyte 2006] Bailey, T., Durrant-Whyte, H.: Simultaneous localization and mapping (SLAM): Part II. IEEE Robotics & Automation Magazine 13(3), 108–117 (2006)

[Bailey *et al.* 2006] Bailey, T., Nieto, J., Guivant, J., Stevens, M., Nebot, E.: Consistency of the EKF-SLAM algorithm. In: IEEE/RSJ International Conference on Intelligent Robots and Systems, pp. 3562–3568 (2006)

[Bailey 2003] Bailey, T.: Constrained initialisation for bearing-only SLAM. In: Proceedings of the IEEE International Conference on Robotics and Automation, vol. 2, pp. 1966–1971 (2003)

[Baker *et al.* 2007] Baker, S., Scharstein, D., Lewis, J.P., Roth, S., Black, M.J., Szeliski, R.: A database and evaluation methodology for optical flow. International Journal of Computer Vision 92(1), 1–31 (2007)

[Bao & Savarese 2011] Bao, S.Y., Savarese, S.: Semantic Structure from Motion. In: IEEE Computer Vision and Pattern Recognition (2011)

[Bar-Shalom & Fortmann 1988] Bar-Shalom, Y., Fortmann, T.E.: Tracking and data association. In: Mathematics in Science and Engineering, vol. 179. Academic Press, INC., San Diego (1988)

[Bar-Shalom *et al.* 2001] Bar-Shalom, Y., Li, X.R., Kirubarajan, T., Wiley, J.: Estimation with applications to tracking and navigation. Wiley Online Library (2001)

[Bay *et al.* 2008] Bay, H., Ess, A., Tuytelaars, T., Van Gool, L.: SURF: Speeded Up Robust Features. Computer Vision and Image Understanding 110(3), 346–359 (2008)

[Blanco *et al.* 2009] Blanco, J.-L., Moreno, F.-A., González, J.: A Collection of Outdoor Robotic Datasets with centimeter-accuracy Ground Truth. Autonomous Robots 27(4), 327–351 (2009)

[Blom & Bar-Shalom 1988] Blom, H.A.P., Bar-Shalom, Y.: The interacting multiple model algorithm for systems with Markovian switching coefficients. IEEE Transactions on Automatic Control 33(8), 780–783 (1988)

[Borenstein *et al.* 1996] Borenstein, J., Everett, H., Feng, L.: Where am I? Sensors and methods for mobile robot positioning, vol. 119, p. 120. University of Michigan (1996)

[Broida *et al.* 1990] Broida, T.J., Chandrashekhar, S., Chellappa, R.: Recursive 3-D motion estimation from a monocular image sequence. IEEE Transactions on Aerospace and Electronic Systems 26(4), 639–656 (1990)

[Brown *et al.* 2003] Brown, M.Z., Burschka, D., Hager, G.D.: Advances in computational stereo. IEEE Transactions on Pattern Analysis and Machine Intelligence, 993–1008 (2003)

[Brown 2003] Brown, D.G., Lowe, M.: Recognising panoramas. In: International Conference on Computer Vision, Nice, pp. 1218–1225 (2003)

[Bryson & Sukkarieh 2005] Bryson, M., Sukkarieh, S.: Bearing-Only SLAM for an Airborne Vehicle. In: Australian Conference on Robotics and Automation (ACRA 2005), Sidney (2005)

[Bryson *et al.* 2009] Bryson, M., Johnson-Roberson, M., Sukkarieh, S.: Airborne Smoothing and Mapping using Vision and Inertial Sensors. In: Proceedings of the IEEE International Conference on Robotics and Automation, pp. 2037–2042 (2009)

[Buckingham 1914] Buckingham, E.: On Physically Similar Systems; Illustrations of the Use of Dimensional Equations. Physical Review 4(4), 345–376 (1914)

[Canavos 1984] Canavos, G.C.: Applied probability and statistical methods. Little, Brown and Company, Boston (1984)

[Canny 1986] Canny, J.: A computational approach to edge detection. IEEE Transactions on Pattern Analysis and Machine Intelligence 8(6), 679–698 (1986)

[Capel & Zisserman 1998] Capel, D., Zisserman, A.: Automated Mosaicing with Super-resolution Zoom. In: Proceedings of the IEEE Conference on Computer Vision and Pattern Recognition, pp. 885–891 (1998)

[Capel 2005] Capel, D.: An effective bail-out test for RANSAC consensus scoring. In: Proceedings of the British Machine Vision Conference, pp. 629–638 (2005)

[Castellanos & Tardós 1999] Castellanos, J.A., Tardós, J.D.: Mobile robot localization and map building: A multisensor fusion approach. Kluwer Academic Publishers, Boston (1999)

[Castellanos *et al.* 1994] Castellanos, J.A., Montiel, J.M.M., Neira, J., Tardós, J.D.: Sensor Influence in the Performance of Simultaneous Mobile Robot Localization and Map Building. LNCIS, vol. 250, pp. 287–296 (1994)

[Castellanos *et al.* 1999] Castellanos, J.A., Montiel, J.M.M., Neira, J., Tardós, J.D.: The SPmap: a probabilistic framework for simultaneous localizationand map building. IEEE Transactions on Robotics and Automation 15(5), 948–952 (1999)

[Castellanos *et al.* 2001] Castellanos, J.A., Neira, J., Tardós, J.D.: Multisensor fusion for simultaneous localization and map building. IEEE Transactions on Robotics and Automation 17(6), 908–914 (2001)

[Castellanos *et al.* 2004] Castellanos, J.A., Neira, J., Tardós, J.D.: Limits to the consistency of EKF-based SLAM. In: 5th IFAC Symposium on Intelligent Autonomous Vehicles (2004)

[Castellanos *et al.* 2007] Castellanos, J.A., Martinez-Cantin, R., Tardós, J.D., Neira, J.: Robocentric map joining: Improving the consistency of EKF-SLAM. Robotics and Autonomous Systems 55(1), 21–29 (2007)

[Castle *et al.* 2010] Castle, R.O., Klein, G., Murray, D.W.: Combining monoSLAM with object recognition for scene augmentation using a wearable camera. Image and Vision Computing 28(11), 1548–1556 (2010)

[Chekhlov *et al.* 2007] Chekhlov, D., Pupilli, M., Mayol, W.W., Calway, A.: Robust Real-Time Visual SLAM Using Scale Prediction and Exemplar Based Feature Description. In: Proceedings of the IEEE Computer Vision and Pattern Recognition (2007)

[Cheng *et al.* 2006] Cheng, Y., Maimone, M., Matthies, L.: Visual odometry on the Mars exploration rovers-a tool to ensure accurate driving and science imaging. IEEE Robotics and Automation Magazine 13(2), 54–62 (2006)

[Chiuso *et al.* 2002] Chiuso, A., Favaro, P., Jin, H., Soatto, S.: Structure from motion causally integrated over time. IEEE Transactions on Pattern Analysis and Machine Intelligence 24(4), 523–535 (2002)

[Chli & Davison 2008] Chli, M., Davison, A.J.: Active matching. In: Forsyth, D., Torr, P., Zisserman, A. (eds.) ECCV 2008, Part I. LNCS, vol. 5302, pp. 72–85. Springer, Heidelberg (2008)

[Chowdhury & Chellappa 2003] Chowdhury, A.K.R., Chellappa, R.: Stochastic Approximation and Rate-Distortion Analysis for Robust Structure and Motion Estimation. International Journal of Computer Vision 55(1), 27–53 (2003)

[Chum & Matas 2008] Chum, O., Matas, J.: Optimal randomized RANSAC. IEEE Transactions on Pattern Analysis and Machine Intelligence 30(8), 1472–1482 (2008)

[Civera *et al.* 2007a] Civera, J., Davison, A.J., Montiel, J.M.M.: Inverse Depth to Depth Conversion for Monocular SLAM. In: IEEE International Conference on Robotics and Automation, April 2007, pp. 2778–2783 (2007)

[Civera *et al.* 2007b] Civera, J., Davison, A.J., Montiel, J.M.M.: Dimensionless monocular SLAM. In: Martí, J., Benedí, J.M., Mendonça, A.M., Serrat, J. (eds.) IbPRIA 2007. LNCS, vol. 4478, pp. 412–419. Springer, Heidelberg (2007)

[Civera *et al.* 2008a] Civera, J., Davison, A.J., Montiel, J.M.M.: Inverse Depth Parametrization for Monocular SLAM. IEEE Transactions on Robotics 24(5), 932–945 (2008)

[Civera *et al.* 2008b] Civera, J., Davison, A.J., Montiel, J.M.M.: Interacting multiple model monocular SLAM. In: IEEE International Conference on Robotics and Automation, pp. 3704–3709 (2008)

[Civera *et al.* 2009a] Civera, J., Bueno, D.R., Davison, A.J., Montiel, J.M.M.: Camera Self-Calibration for Sequential Bayesian Structure From Motion. In: IEEE International Conference on Robotics and Automation, ICRA 2009 (2009)

[Civera *et al.* 2009b] Civera, J., Davison, A.J., Magallon, J.A., Montiel, J.M.M.: Drift-Free Real-Time Sequential Mosaicing. International Journal of Computer Vision 81(2), 128–137 (2009)

[Civera *et al.* 2009c] Civera, J., Grasa, O.G., Davison, A.J., Montiel, J.M.M.: 1-Point RANSAC for EKF-Based Structure from Motion. In: Proceedings of the IEEE/RSJ International Conference on Intelligent Robots and Systems, pp. 3498–3504 (2009)

[Civera *et al.* 2010] Civera, J., Grasa, O.G., Davison, A.J., Montiel, J.M.M.: 1-Point RANSAC for EKF Filtering: Application to Real-Time Structure from Motion and Visual Odometry. Journal of Field Robotics 27(5), 609–631 (2010)

[Civera *et al.* 2011] Civera, J., Gálvez-López, D., Riazuelo, L., Tardós, J.D., Montiel, J.M.M.: Towards Semantic SLAM using a Monocular Camera. In: Proceedings of the IEEE/RSJ International Conference on Intelligent Robots and Systems (2011)

[Clemente *et al.* 2007] Clemente, L.A., Davison, A.J., Reid, I.D., Neira, J., Tardos, J.D.: Mapping Large Loops with a Single Hand-Held Camera. In: Proceedings of Robotics: Science and Systems (2007)

[Comport *et al.* 2007] Comport, A.I., Malis, E., Rives, P.: Accurate Quadrifocal Tracking for Robust 3D Visual Odometry. In: 2007 IEEE International Conference on Robotics and Automation, April 2007, pp. 40–45 (2007)

[Davison & Murray 1998] Davison, A.J., Murray, D.W.: Mobile robot localisation using active vision. In: Burkhardt, H., Neumann, B. (eds.) ECCV 1998. LNCS, vol. 1407, pp. 809–825. Springer, Heidelberg (1998)

[Davison *et al.* 2007] Davison, A.J., Molton, N.D., Reid, I.D., Stasse, O.: MonoSLAM: Real-Time Single Camera SLAM. IEEE Transactions on Pattern Analysis and Machine Intelligence, 1052–1067 (June 2007)

[Davison 2003] Davison, A.J.: Real-time simultaneous localisation and mapping with a single camera. In: Proceedings of the Ninth IEEE International Conference on Computer Vision, pp. 1403–1410 (2003)

[de Agapito *et al.* 2001] de Agapito, L., Hayman, E., Reid, I.A.: Self-Calibration of Rotating and Zooming Cameras. International Journal of Computer Vision 45(2), 107–127 (2001)

[Dissanayake *et al.* 2001] Dissanayake, M., Newman, P., Clark, S., Durrant-Whyte, H.F., Csorba, M.: A solution to the simultaneous localization and map building (SLAM) problem. IEEE Transactions on Robotics and Automation 17(3), 229–241 (2001)

[Durrant-Whyte & Bailey 2006] Durrant-Whyte, H., Bailey, T.: Simultaneous localisation and mapping (SLAM): Part I the essential algorithms. Robotics and Automation Magazine 13(2), 99–110 (2006)

[Durrant-Whyte *et al.* 2003] Durrant-Whyte, H., Majumder, S., Thrun, S., de Battista, M., Scheding, S.: A Bayesian Algorithm for Simultaneous Localisation and Map Building. In: The 10th International Symposium on Robotics Research, pp. 49–60 (2003)

[Eade & Drummond 2006] Eade, E., Drummond, T.: Scalable Monocular SLAM. In: Proceedings of the IEEE Conference on Computer Vision and Pattern Recognition, New York, pp. 469–476 (2006)

[Eade & Drummond 2007] Eade, E., Drummond, T.: Monocular slam as a graph of coalesced observations. In: IEEE 11th International Conference on Computer Vision, ICCV 2007, pp. 1–8 (2007)

[Eustice *et al.* 2005] Eustice, R.M., Singh, H., Leonard, J.J., Walter, M., Ballard, R.: Visually Navigating the RMS Titanic with SLAM Information Filters. In: Proceedings of Robotics: Science and Systems (2005)

[Everingham *et al.* 2010] Everingham, M., Van Gool, L., Williams, C.K.I., Winn, J., Zisserman, A.: The PASCAL visual object classes (VOC) challenge. International Journal of Computer Vision 88(2), 303–338 (2010)

[Faugeras *et al.* 1992] Faugeras, O.D., Luong, Q.T., Maybank, S.J.: Camera self-calibration: theory and experiments. In: Sandini, G. (ed.) ECCV 1992. LNCS, vol. 588, pp. 321–334. Springer, Heidelberg (1992)

[Feder *et al.* 1999] Feder, H.J.S., Leonard, J.J., Smith, C.M.: Adaptive Mobile Robot Navigation and Mapping. International Journal of Robotics Research 18(7), 650–668 (1999)

[Fenwick *et al.* 2002] Fenwick, J.W., Newman, P.M., Leonard, J.J.: Cooperative concurrent mapping and localization. In: Proceedings of the 2002 IEEE International Conference on Robotics and Automation, vol. 2, pp. 1810–1817 (2002)

[Fischler & Bolles 1981] Fischler, M.A., Bolles, R.C.: Random Sample Consensus, a Paradigm for Model Fitting with Applications to Image Analysis and Automated Cartography. Communications of the ACM 24(6), 381–395 (1981)

[Fitzgibbon & Zisserman 1998] Fitzgibbon, A.W., Zisserman, A.: Automatic Camera Recovery for Closed or Open Image Sequences. In: Burkhardt, H.-J., Neumann, B. (eds.) ECCV 1998. LNCS, vol. 1406, pp. 311–326. Springer, Heidelberg (1998)

[Foxlin 2002] Foxlin, E.: Generalized architecture for simultaneous localization, autocalibration and map-building. In: Proceedings of the IEEE/RSJ Conference on Intelligent Robots and Systems, pp. 2–4 (2002)

[Frahm & Pollefeys 2006] Frahm, J.M., Pollefeys, M.: RANSAC for (quasi-) degenerate data (QDEGSAC). In: 2006 IEEE Conference on Computer Vision and Pattern Recognition, vol. 1, pp. 453–460 (2006)

[Funke & Pietzsch 2009] Funke, J., Pietzsch, T.: A Framework For Evaluating Visual SLAM. In: Proceedings of the British Machine Vision Conference (2009)

[Gelb 1999] Gelb, A.: Applied optimal estimation. MIT Press (1999)

[Grasa *et al.* 2009a] Grasa, O.G., Civera, J., Guemes, A., Munöz, V., Montiel, J.M.M.: EKF Monocular SLAM 3D Modeling, Measuring and Augmented Reality from Endoscope Image Sequences. In: 5th Workshop on Augmented Environments for Medical Imaging including Augmented Reality in Computer-Aided Surgery, held in conjunction with MICCAI 2009 (2009)

[Grasa *et al.* 2009b] Grasa, O.G., Civera, J., Guemes, A., Munöz, V., Montiel, J.M.M.: Real-time 3D Modeling from Endoscope Image Sequences. In: Proceedings of the 2009 IEEE International Conference on Robotics and Automation; Workshop on Advanced Sensing and Sensor Integration in Medical Robotics (2009)

[Grasa *et al.* 2011] Grasa, O.G., Civera, J., Montiel, J.M.M.: EKF monocular SLAM with relocalization for laparoscopic sequences. In: Proceedings of the IEEE International Conference on Robotics and Automation (2011)

[Handa *et al.* 2010] Handa, A., Chli, M., Strasdat, H., Davison, A.J.: Scalable Active Matching. In: Proceedings of the IEEE Conference on Computer Vision and Pattern Recognition, pp. 1546–1553 (2010)

[Harris & Stephens 1988] Harris, C.G., Stephens, M.: A Combined Corner and Edge Detector. In: Proceedings of the 4th Alvey Vision Conference, pp. 147–151 (1988)

[Hartley & Zisserman 2004] Hartley, R.I., Zisserman, A.: Multiple view geometry in computer vision. Cambridge University Press (2004), ISBN: 0521540518

[Heeger & Jepson 1992] Heeger, D.J., Jepson, A.D.: Subspace methods for recovering rigid motion I: Algorithm and implementation. International Journal of Computer Vision 7(2), 95–117 (1992)

[Holmes *et al.* 2008] Holmes, S., Klein, G., Murray, D.: An O (N2) Square Root Unscented Kalman Filter for Visual Simultaneous Localization and Mapping. IEEE Transactions on Pattern Analysis and Machine Intelligence 22, 1251–1263 (2008)

[Jaynes 2003] Jaynes, E.T.: Probability theory: The logic of science. Cambridge University Press (2003)

[Julier & Uhlmann 2001] Julier, S.J., Uhlmann, J.K.: A counter example to the theory of simultaneous localization and map building. In: IEEE International Conference on Robotics and Automation, vol. 4, pp. 4238–4243 (2001)

[Jung & Lacroix 2003] Jung, I.K., Lacroix, S.: High resolution terrain mapping using low altitude aerial stereo imagery. In: Proceedings of the Ninth International Conference on Computer Vision, p. 946 (2003)

[Kalman 1960] Kalman, R.E.: A new approach to linear filtering and prediction problems. Journal of basic Engineering 82(1), 35–45 (1960)

[Kanatani 2004] Kanatani, K.: Uncertainty modeling and model selection for geometric inference. IEEE Transactions on Pattern Analysis and Machine Intelligence 26(10), 1307–1319 (2004)

[Kim & Hong 2006] Kim, D.W., Hong, K.S.: Real-time mosaic using sequential graph. Journal of Electronic Imaging 15(2), 47–63 (2006)

[Kim & Sukkarieh 2003] Kim, J.H., Sukkarieh, S.: Airborne Simultaneous Localisation and Map Building. In: Proceedings of the IEEE International Conference on Robotics and Automation, pp. 406–411 (2003)

[Kirubarajan *et al.* 1998] Kirubarajan, T., Bar-Shalom, Y., Blair, W.D., Watson, G.A.: IMMPDAF for Radar Management and Tracking Benchmark with ECM. IEEE Transactions on Aerospace and Electronic Systems 34(4), 1115–1134 (1998)

[Klein & Murray 2007] Klein, G., Murray, D.: Parallel Tracking and Mapping for Small AR Workspaces. In: Proceedings of the Sixth IEEE and ACM International Symposium on Mixed and Augmented Reality, pp. 1–10 (2007)

[Klein & Murray 2008] Klein, G., Murray, D.: Improving the Agility of Keyframe-Based SLAM. In: Forsyth, D., Torr, P., Zisserman, A. (eds.) ECCV 2008, Part II. LNCS, vol. 5303, pp. 802–815. Springer, Heidelberg (2008)

[Klein & Murray 2009] Klein, G., Murray, D.: Parallel tracking and mapping on a camera phone. In: Proceedings of the 8th IEEE and ACM International Symposium on Mixed and Augmented Reality, pp. 83–86 (2009)

[Konolige & Agrawal 2008] Konolige, K., Agrawal, M.: FrameSLAM: From bundle adjustment to realtime visual mappping. IEEE Transactions on Robotics 24(5), 1066–1077 (2008)

[Konolige *et al.* 2007] Konolige, K., Agrawal, M., Solà, J.: Large-Scale Visual Odometry for Rough Terrain. In: Proceedings of the International Symposium on Research in Robotics, pp. 201–212 (2007)

[Kummerle *et al.* 2009] Kummerle, R., Steder, B., Dornhege, C., Ruhnke, M., Grisetti, G., Stachniss, C., Kleiner, A.: On measuring the accuracy of SLAM algorithms. Autonomous Robots 27(4), 387–407 (2009)

[Kwok & Dissanayake 2004] Kwok, N.M., Dissanayake, G.: An Efficient Multiple Hypothesis Filter for Bearing-Only SLAM. In: IEEE/RSJ International Conference on Intelligent Robots and Systems, pp. 736–741 (2004)

[Kwok *et al.* 2005] Kwok, N.M., Dissanayake, G., Ha, Q.P.: Bearing-only SLAM Using a SPRT Based Gaussian Sum Filter. In: Proceedings of the 2005 IEEE International Conference on Robotics and Automation, pp. 1109–1114 (2005)

[Li & Jilkov 2005] Rong Li, X., Jilkov, V.P.: Survey of Maneuvering Target Tracking. Part V: Multiple-Model Methods. IEEE Transactions on Aerospace and Electronic Systems 41(4), 1255–1320 (2005)

[Lovegrove & Davison 2010] Lovegrove, S., Davison, A.: Real-time spherical mosaicing using whole image alignment. In: 11th European Conference on Computer Vision, pp. 73–86 (2010)

[Lowe 2004] Lowe, D.G.: Distinctive image features from scale-invariant keypoints. International Journal of Computer Vision 60(2), 91–110 (2004)

[Marks *et al.* 1995] Marks, R.L., Rock, S.M., Lee, M.J.: Real-time Video Mosaicking of the Ocean Floor. IEEE Journal of Oceanic Engineering 20(3), 229–241 (1995)

[Marzorati *et al.* 2008] Marzorati, D., Matteucci, M., Migliore, D., Sorrenti, D.G.: Monocular slam with inverse scaling parametrization. In: Proceedings of 2008 British Machine Vision Conference (BMVC 2008), pp. 945–954. Citeseer (2008)

[Matthies *et al.* 1989] Matthies, L., Kanade, T., Szeliski, R.: Kalman Filter-based Algorithms for Estimating Depth from Image Sequences. International Journal of Computer Vision 3(3), 209–238 (1989)

[Maybank & Faugeras 1992] Maybank, S.J., Faugeras, O.: A Theory of Self-Calibration of a Moving Camera. International Journal of Computer Vision 8(2), 123–151 (1992)

[Mikhail *et al.* 2001] Mikhail, E.M., Bethel, J.S., McGlone, J.C.: Introduction to modern photogrammetry. John Wiley & Sons (2001)

[Mikolajczyk & Schmid 2005] Mikolajczyk, K., Schmid, C.: A performance evaluation of local descriptors. IEEE Transactions on Pattern Analysis and Machine Intelligence 27(10), 1615–1630 (2005)

[Mikolajczyk *et al.* 2005] Mikolajczyk, K., Tuytelaars, T., Schmid, C., Zisserman, A., Matas, J., Schaffalitzky, F., Kadir, T., Gool, L.V.: A comparison of affine region detectors. International Journal of Computer Vision 65(1), 43–72 (2005)

[Molton *et al.* 2004] Molton, N.D., Davison, A.J., Reid, I.D.: Locally Planar Patch Features for Real-Time Structure from Motion. In: Proceedings of the 15th British Machine Vision Conference, Kingston (2004)

[Montiel *et al.* 2006] Montiel, J.M.M., Civera, J., Davison, A.J.: Unified Inverse Depth Parametrization for Monocular SLAM. In: Proceedings of Robotics: Science and Systems, Philadelphia, USA (August 2006)

[Moreno-Noguer *et al.* 2008] Moreno-Noguer, F., Lepetit, V., Fua, P.: Pose priors for simultaneously solving alignment and correspondence. In: Forsyth, D., Torr, P., Zisserman, A. (eds.) ECCV 2008, Part II. LNCS, vol. 5303, pp. 405–418. Springer, Heidelberg (2008)

[Morimoto & Chellappa 1997] Morimoto, C., Chellappa, R.: Fast 3D Stabilization and Mosaic Construction. In: Proceedings of the IEEE Computer Vision and Pattern Recognition, pp. 660–665 (1997)

[Mouragnon *et al.* 2006] Mouragnon, E., Lhuillier, M., Dhome, M., Dekeyser, F., Sayd, P.: Real-Time Localization and 3D Reconstruction. In: Proceedings of the IEEE Computer Vision and Pattern Recognition, vol. 1, pp. 363–370 (2006)

[Mouragnon *et al.* 2009] Mouragnon, E., Lhuillier, M., Dhome, M., Dekeyser, F., Sayd, P.: Generic and real-time structure from motion using local bundle adjustment. Image and Vision Computing 27(8), 1178–1193 (2009)

[Neira & Tardós 2001] Neira, J., Tardós, J.D.: Data Association in Stochastic Mapping using the Joint Compatibility Test. IEEE Transactions on Robotics and Automation 17(6), 890–897 (2001)

[Newman *et al.* 2002] Newman, P.M., Leonard, J.J., Neira, J., Tardós, J.: Explore and Return: Experimental Validation of Real Time Concurrent Mapping and Localization. In: Proceedings of the IEEE International Conference on Robotics and Automation, pp. 1802–1809 (2002)

[Nistér *et al.* 2004] Nistér, D., Naroditsky, O., Bergen, J.: Visual Odometry. In: Proceedings of the IEEE Conference on Computer Vision and Pattern Recognition, vol. 1, pp. 652–659 (2004)

[Nistér *et al.* 2006] Nistér, D., Naroditsky, O., Bergen, J.: Visual odometry for ground vehicle applications. Journal of Field Robotics 23(1), 3–20 (2006)

[Nistér 2004] Nistér, D.: An efficient solution to the five-point relative pose problem. IEEE Transactions on Pattern Analysis and Machine Intelligence 26(6), 756–770 (2004)

[Nistér 2005] Nistér, D.: Preemptive RANSAC for live structure and motion estimation. Machine Vision and Applications 16(5), 321–329 (2005)

[Okutomi & Kanade 1993] Okutomi, M., Kanade, T.: A Multiple-Baseline Stereo. IEEE Transactions on Pattern Analysis and Machine Intelligence 15(4), 353–363 (1993)

[Ortín & Montiel 2001] Ortín, D., Montiel, J.M.M.: Indoor robot motion based on monocular images. Robotica 19(03), 331–342 (2001)

[Ortín *et al.* 2003] Ortín, D., Montiel, J.M.M., Zisserman, A.: Automated Multisensor Polyhedral Model Acquisition. In: Proceedings of the IEEE International Conference on Robotics and Automation (2003)

[Paz *et al.* 2008a] Paz, L.M., Piniés, P., Tardós, J.D., Neira, J.: Large-Scale 6-DOF SLAM With Stereo-in-Hand. IEEE Transactions on Robotics 24(5), 946–957 (2008)

[Paz *et al.* 2008b] Paz, L.M., Tardós, J.D., Neira, J.: Divide and Conquer: EKF SLAM in O (n). IEEE Transactions on Robotics 24(5), 1107–1120 (2008)

[PhotoTourism 2011] PhotoTourism (June 2011), http://phototour.cs.washington.edu/

[Piniés & Tardós 2008] Piniés, P., Tardós, J.D.: Large Scale SLAM Building Conditionally Independent Local Maps: Application to Monocular Vision. IEEE Transactions on Robotics 24(5) (2008)

[Piniés *et al.* 2006] Piniés, P., Tardós, J.D., Neira, J.: Localization of avalanche victims using robocentric SLAM. In: Proc. IEEE/RSJ International Conference on Intelligent Robots and Systems, pp. 3074–3079 (2006)

[Piniés *et al.* 2007] Piniés, P., Lupton, T., Sukkarieh, S., Tardós, J.D.: Inertial aiding of inverse depth SLAM using a monocular camera. In: Proceedings of the 2007 IEEE International Conference on Robotics and Automation, pp. 2797–2802 (2007)

[Pollefeys *et al.* 1999] Pollefeys, M., Koch, R., Van Gool, L.: Self-Calibration and Metric Reconstruction Inspite of Varying and Unknown Intrinsic Camera Parameters. International Journal of Computer Vision 32(1), 7–25 (1999)

[Qian & Chellappa 2004] Qian, G., Chellappa, R.: Bayesian self-calibration of a moving camera. Computer Vision and Image Understanding 95(3), 287–316 (2004)

[Raguram *et al.* 2008] Raguram, R., Frahm, J.-M., Pollefeys, M.: A comparative analysis of RANSAC techniques leading to adaptive real-time random sample consensus. In: Forsyth, D., Torr, P., Zisserman, A. (eds.) ECCV 2008, Part II. LNCS, vol. 5303, pp. 500–513. Springer, Heidelberg (2008)

[RAWSEEDS 2011] RAWSEEDS. RAWSEEDS public datasets web page (June 2011), http://www.rawseeds.org/

[Renka 1997] Renka, R.J.: Algorithm 772: STRIPACK: Delaunay Triangulation and Voronoi Diagram on the Surface of a Sphere. ACM Transactions on Mathematical Software 23(3), 416–434 (1997)

[Rituerto *et al.* 2010] Rituerto, A., Puig, L., Guerrero, J.J.: Visual SLAM with an omnidirectional camera. In: 2010 International Conference on Pattern Recognition, pp. 348–351 (2010)

[Rosten & Drummond 2005] Rosten, E., Drummond, T.: Fusing points and lines for high performance tracking. In: Proceedings of the 10th IEEE International Conference on Computer Vision, vol. 2, pp. 1508–1515 (2005)

[Rousseeuw & Leroy 1987] Rousseeuw, P.J., Leroy, A.M.: Robust regression and outlier detection. Wiley Interscience, New York (1987)

[Sawhney *et al.* 1998] Sawhney, H.S., Hsu, S., Kumar, R.: Robust Video Mosaicing through Topology Inference and Local to Global Alignment. In: Burkhardt, H., Neumann, B. (eds.) ECCV 1998. LNCS, vol. 1407, pp. 103–119. Springer, Heidelberg (1998)

[Scaramuzza *et al.* 2007] Scaramuzza, D., Harati, A., Siegwart, R.: Extrinsic self calibration of a camera and a 3d laser range finder from natural scenes. In: Proceedings of the IEEE/RSJ International Conference on Intelligent Robots and Systems, pp. 4164–4169 (2007)

[Scaramuzza *et al.* 2009] Scaramuzza, D., Fraundorfer, F., Siegwart, R.: Real-Time Monocular Visual Odometry for On-Road Vehicles with 1-Point RANSAC. In: Proceedings of the IEEE International Conference on Robotics and Automation, pp. 4293–4299 (2009)

[Scharstein & Szeliski 2002] Scharstein, D., Szeliski, R.: A taxonomy and evaluation of dense two-frame stereo correspondence algorithms. International Journal of Computer Vision 47(1), 7–42 (2002)

[Schindler & Suter 2006] Schindler, K., Suter, D.: Two-view multibody structure-and-motion with outliers through model selection. IEEE Transactions on Pattern Analysis and Machine Intelligence 28(6), 983–995 (2006)

[Sibley *et al.* 2009] Sibley, G., Mei, C., Reid, I., Newman, P.: Adaptive Relative Bundle Adjustment. In: Proceedings of Robotics: Science and Systems, Seattle, USA, June 2009, pp. 976–982 (2009)

[Smith & Cheeseman 1986a] Smith, R.C., Cheeseman, P.: On the Representation and Estimation of Spatial Uncertainty. International Journal of Robotics Research 5(4), 56–68 (1986)

[Smith & Cheeseman 1986b] Smith, R.C., Cheeseman, P.: On the representation and estimation of spatial uncertainty. The International Journal of Robotics Research 5(4), 56 (1986)

[Smith *et al.* 1987] Smith, R., Self, M., Cheeseman, P.: A stochastic map for uncertain spatial relationships. In: 4th International Symposium on Robotics Research (1987)

[Smith *et al.* 2009] Smith, M., Baldwin, I., Churchill, W., Paul, R., Newman, P.: The New College Vision and Laser Data Set. The International Journal of Robotics Research 28(5), 595–599 (2009)

[Solà *et al.* 2005] Solà, J., Monin, A., Devy, M., Lemaire, T.: Undelayed Initialization in Bearing Only SLAM. In: 2005 IEEE/RSJ International Conference on Intelligent Robots and Systems (2005)

[Solà *et al.* 2008] Solà, J., Monin, A., Devy, M., Vidal-Calleja, T.: Fusing Monocular Information in Multicamera SLAM. IEEE Transactions on Robotics 24(5), 958–968 (2008)

[Solà *et al.* 2009] Solà, J., Vidal-Calleja, T., Devy, M.: Undelayed initialization of line segments in monocular SLAM. In: IEEE/RSJ International Conference on Intelligent Robots and Systems, IROS 2009, pp. 1553–1558. IEEE (2009)

[Solà *et al.* 2011] Solà, J., Vidal-Calleja, T., Civera, J., Montiel, J.M.M.: Impact of landmark parametrization on monocular EKF-SLAM with points and lines. Accepted for publication in International Journal of Computer Vision (2011)

[Solà 2007] Solà, J.: Towards Visual Localization, Mapping and Moving Objects Tracking by a Mobile Robot: a Geometric and Probabilistic Approach. PhD thesis, LAAS-CNRS (2007)

[Solà 2010] Solà, J.: Consistency of the monocular EKF-SLAM algorithm for three different landmark parametrizations. In: Proceedings of the 2010 IEEE International Conference on Robotics and Automation, pp. 3513–3518 (2010)

[Steedly *et al.* 2005] Steedly, D., Pal, C., Szeliski, R.: Efficiently registering video into panoramic mosaics. In: Proceedings of the 10th IEEE International Conference on Computer Vision, vol. 2, pp. 1300–1307 (2005)

[Strasdat *et al.* 2010a] Strasdat, H., Montiel, J.M.M., Davison, A.J.: Real-time monocular SLAM: Why filter? In: IEEE International Conference on Robotics and Automation (ICRA), pp. 2657–2664 (2010)

[Strasdat *et al.* 2010b] Strasdat, H., Montiel, J.M.M., Davison, A.J.: Scale drift-aware large scale monocular SLAM. In: Proceedings of Robotics: Science and Systems (RSS) (2010)

[Strasdat *et al.* 2011] Strasdat, H., Davison, A.J., Montiel, J.M.M., Konolige, K.: Double Window Optimisation for Constant Time Visual SLAM. In: 13th International Conference on Computer Vision (2011)

[Szeliski & Shum 1997] Szeliski, R., Shum, H.Y.: Creating full view panoramic image mosaics and environment maps. In: Proceedings of the 24th Annual Conference on Computer Graphics and Interactive Techniques, pp. 251–258 (1997)

[Tardif *et al.* 2008] Tardif, J.P., Pavlidis, Y., Daniilidis, K.: Monocular visual odometry in urban environments using an omnidirectional camera. In: Proceedings of the IEEE/RSJ International Conference on Intelligent Robots and Systems, pp. 2531–2538 (2008)

[Tardós *et al.* 2002] Tardós, J.D., Neira, J., Newman, P.M., Leonard, J.J.: Robust mapping and localization in indoor environments using sonar data. The International Journal of Robotics Research 21(4), 311 (2002)

[Thrun *et al.* 2005] Thrun, S., Burgard, W., Fox, D.: Probabilistic robotics. MIT Press, Cambridge (2005)

[Torr & Murray 1993] Torr, P.H.S., Murray, D.W.: Outlier detection and motion segmentation. Sensor Fusion VI 2059, 432–443 (1993)

[Torr & Zisserman 2000] Torr, P.H.S., Zisserman, A.: MLESAC: A new robust estimator with application to estimating image geometry. Computer Vision and Image Understanding 78(1), 138–156 (2000)

[Torr *et al.* 1999] Torr, P.H.S., Fitzgibbon, A., Zisserman, A.: The Problem of Degeneracy in Structure and Motion Recovery from Uncalibrated Image Sequences. International Journal of Computer Vision 32(1), 27–45 (1999)

[Torr 2002] Torr, P.H.S.: Bayesian Model Estimation and Selection for Epipolar Geometry and Generic Manifold Fitting. International Journal of Computer Vision 50(1), 35–61 (2002)

[Trawny & Roumeliotis 2006] Trawny, N., Roumeliotis, S.I.: A unified framework for nearby and distant landmarks in bearing-only SLAM. In: Proceedings of the IEEE International Conference on Robotics and Automation, pp. 1923–1929 (2006)

[Triggs *et al.* 2000] Triggs, B., McLauchlan, P.F., Hartley, R.I., Fitzgibbon, A.W.: Bundle Adjustment – A Modern Synthesis. In: Triggs, B., Zisserman, A., Szeliski, R. (eds.) ICCV-WS 1999. LNCS, vol. 1883, pp. 298–375. Springer, Heidelberg (2000)

[Vedaldi *et al.* 2005] Vedaldi, A., Jin, H., Favaro, P., Soatto, S.: KALMANSAC: Robust filtering by consensus. In: 10th IEEE International Conference on Computer Vision, vol. 1, pp. 633–640 (2005)

[Wald 1945] Wald, A.: Sequential Tests of Statistical Hypothesis. The Annals of Mathematical Statistics 16(2), 117–186 (1945)

[Williams *et al.* 2007] Williams, B., Klein, G., Reid, I.: Real-time SLAM relocalisation. In: IEEE 11th International Conference on Computer Vision, pp. 1–8 (2007)

[Williams *et al.* 2008] Williams, B., Cummins, M., Neira, J., Newman, P., Reid, I., Tardos, J.: An image–to–map loop closing method for monocular SLAM. In: Proceedings of the IEEE/RSJ International Conference on Intelligent Robots and Systems, pp. 2053–2059 (2008)

[Williams 2009] Williams, B.: Simultaneous Localisation and Mapping Using a Single Camera. PhD thesis, University of Oxford (2009)

[Yang & Toderici 2011] Yang, W., Toderici, G.: Discriminative Tag Learning on YouTube Videos with Latent Sub-tags. In: IEEE Computer Vision and Pattern Recognition (2011)

[Zhu *et al.* 2006] Zhu, Z., Xu, G., Riseman, E.M., Hanson, A.R.: Fast Construction of Dynamic and Multi-Resolution 360° Panoramas from Video Sequences. Image and Vision Computing 24(1), 13–26 (2006)

Springer Tracts in Advanced Robotics

Edited by B. Siciliano, O. Khatib and F. Groen

Further volumes of this series can be found on our homepage: springer.com

Vol. 75: Civera, J.; Davison, A.J.; Montiel, J.M.M.
Structure from Motion Using the Extended Kalman Filter
168 p. 2012 [978-3-642-24833-7]

Vol. 73: Corke, P.;
Robotics, Vision and Control
XXX p. 2011 [978-3-642-20143-1]

Vol. 72: Mullane, J.; Vo, B.-N.; Adams, M.; Vo, B.-T.
Random Finite Sets for Robot Mapping and SLAM
146 p. 2011 [978-3-642-21389-2]

Vol. 70: Pradalier, C.; Siegwart, R.; Hirzinger, G. (Eds.)
Robotics Research
752 p. 2011 [978-3-642-19456-6]

Vol. 69: Rocon, E.; Pons, J.L.
Exoskeletons in Rehabilitation Robotics
138 p. 2010 [978-3-642-17658-6]

Vol. 68: Hsu, D.; Isler, V.; Latombe, J.-C.; Ming C. Lin (Eds.)
Algorithmic Foundations of Robotics IX
424 p. 2010 [978-3-642-17451-3]

Vol. 67: Schütz, D.; Wahl, F.M. (Eds.)
Robotic Systems for Handling and Assembly
460 p. 2010 [978-3-642-16784-3]

Vol. 66: Kaneko, M.; Nakamura, Y. (Eds.)
Robotics Research
450 p. 2010 [978-3-642-14742-5]

Vol. 65: Ribas, D.; Ridao, P.; Neira, J.
Underwater SLAM for Structured Environments Using an Imaging Sonar
142 p. 2010 [978-3-642-14039-6]

Vol. 64: Vasquez Govea, A.D.
Incremental Learning for Motion Prediction of Pedestrians and Vehicles
153 p. 2010 [978-3-642-13641-2]

Vol. 63: Vanderborght, B.;
Dynamic Stabilisation of the Biped Lucy Powered by Actuators with Controllable Stiffness
281 p. 2010 [978-3-642-13416-6]

Vol. 62: Howard, A.; Iagnemma, K.; Kelly, A. (Eds.):
Field and Service Robotics
511 p. 2010 [978-3-642-13407-4]

Vol. 61: Mozos, Ó.M.
Semantic Labeling of Places with Mobile Robots
134 p. 2010 [978-3-642-11209-6]

Vol. 60: Zhu, W.-H.
Virtual Decomposition Control – Toward Hyper Degrees of Freedom Robots
443 p. 2010 [978-3-642-10723-8]

Vol. 59: Otake, M.
Electroactive Polymer Gel Robots – Modelling and Control of Artificial Muscles
238 p. 2010 [978-3-540-23955-0]

Vol. 58: Kröger, T.
On-Line Trajectory Generation in Robotic Systems – Basic Concepts for Instantaneous Reactions to Unforeseen (Sensor) Events
230 p. 2010 [978-3-642-05174-6]

Vol. 57: Chirikjian, G.S.; Choset, H.; Morales, M., Murphey, T. (Eds.)
Algorithmic Foundations of Robotics VIII – Selected Contributions of the Eighth International Workshop on the Algorithmic Foundations of Robotics
680 p. 2010 [978-3-642-00311-0]

Vol. 56: Buehler, M.; Iagnemma, K.; Singh S. (Eds.)
The DARPA Urban Challenge – Autonomous Vehicles in City Traffic
625 p. 2009 [978-3-642-03990-4]

Vol. 55: Stachniss, C.
Robotic Mapping and Exploration
196 p. 2009 [978-3-642-01096-5]

Vol. 54: Khatib, O.; Kumar, V.;
Pappas, G.J. (Eds.)
Experimental Robotics:
The Eleventh International Symposium
579 p. 2009 [978-3-642-00195-6]

Vol. 53: Duindam, V.; Stramigioli, S.
Modeling and Control for Efficient Bipedal
Walking Robots
211 p. 2009 [978-3-540-89917-4]

Vol. 52: Nüchter, A.
3D Robotic Mapping
201 p. 2009 [978-3-540-89883-2]

Vol. 51: Song, D.
Sharing a Vision
186 p. 2009 [978-3-540-88064-6]

Vol. 50: Alterovitz, R.; Goldberg, K.
Motion Planning in Medicine: Optimization
and Simulation Algorithms for
Image-Guided Procedures
153 p. 2008 [978-3-540-69257-7]

Vol. 49: Ott, C.
Cartesian Impedance Control of Redundant
and Flexible-Joint Robots
190 p. 2008 [978-3-540-69253-9]

Vol. 48: Wolter, D.
Spatial Representation and
Reasoning for Robot
Mapping
185 p. 2008 [978-3-540-69011-5]

Vol. 47: Akella, S.; Amato, N.;
Huang, W.; Mishra, B.; (Eds.)
Algorithmic Foundation of Robotics VII
524 p. 2008 [978-3-540-68404-6]

Vol. 46: Bessière, P.; Laugier, C.;
Siegwart R. (Eds.)
Probabilistic Reasoning and Decision
Making in Sensory-Motor Systems
375 p. 2008 [978-3-540-79006-8]

Vol. 45: Bicchi, A.; Buss, M.;
Ernst, M.O.; Peer A. (Eds.)
The Sense of Touch and Its Rendering
281 p. 2008 [978-3-540-79034-1]

Vol. 44: Bruyninckx, H.; Přeučil, L.;
Kulich, M. (Eds.)
European Robotics Symposium 2008
356 p. 2008 [978-3-540-78315-2]

Vol. 43: Lamon, P.
3D-Position Tracking and Control
for All-Terrain Robots
105 p. 2008 [978-3-540-78286-5]

Vol. 42: Laugier, C.; Siegwart, R. (Eds.)
Field and Service Robotics
597 p. 2008 [978-3-540-75403-9]

Vol. 41: Milford, M.J.
Robot Navigation from Nature
194 p. 2008 [978-3-540-77519-5]

Vol. 40: Birglen, L.; Laliberté, T.; Gosselin, C.
Underactuated Robotic Hands
241 p. 2008 [978-3-540-77458-7]

Vol. 39: Khatib, O.; Kumar, V.; Rus, D. (Eds.)
Experimental Robotics
563 p. 2008 [978-3-540-77456-3]

Vol. 38: Jefferies, M.E.; Yeap, W.-K. (Eds.)
Robotics and Cognitive Approaches to
Spatial Mapping
328 p. 2008 [978-3-540-75386-5]

Vol. 37: Ollero, A.; Maza, I. (Eds.)
Multiple Heterogeneous Unmanned Aerial
Vehicles
233 p. 2007 [978-3-540-73957-9]

Vol. 36: Buehler, M.; Iagnemma, K.;
Singh, S. (Eds.)
The 2005 DARPA Grand Challenge – The Great
Robot Race
520 p. 2007 [978-3-540-73428-4]

Vol. 35: Laugier, C.; Chatila, R. (Eds.)
Autonomous Navigation in Dynamic
Environments
169 p. 2007 [978-3-540-73421-5]

Vol. 34: Wisse, M.; van der Linde, R.Q.
Delft Pneumatic Bipeds
136 p. 2007 [978-3-540-72807-8]

Vol. 33: Kong, X.; Gosselin, C.
Type Synthesis of Parallel
Mechanisms
272 p. 2007 [978-3-540-71989-2]

MIX
Papier aus verantwortungsvollen Quellen
Paper from responsible sources
FSC® C105338

FSC
www.fsc.org

If you have any concerns about our products,
you can contact us on
ProductSafety@springernature.com

In case Publisher is established outside the EU,
the EU authorized representative is:
Springer Nature Customer Service Center GmbH
Europaplatz 3, 69115 Heidelberg, Germany

Printed by Libri Plureos GmbH
in Hamburg, Germany